THE NAV/SQL PERFORMANCE FIELD GUIDE

Fixing Trouble with Microsoft® Dynamics™ NAV and Microsoft® SQL Server™

(4. Edition - Version 2009)

STRYK System Improvement
Performance Optimization & Troubleshooting

Imprint

Written by

> Jörg A. Stryk
> STRYK System Improvement
> Germany
>
> http://www.stryk.info/

Copyright © 2007 - 2009 STRYK System Improvement, Thalmässing, Germany
All rights reserved.

Printed and published by Books on Demand GmbH, Norderstedt, Germany
http://www.bod.de/

BoD No.:   510273

ISBN: 978-3-8370-1442-6

1st Edition: October 2007
2nd Edition: December 2007
3rd Edition: October 2008
4th Edition: October 2009

The NAV/SQL Performance Field Guide
Version 2009

Document Version

Version	Date	Description
0.90	August 2006	Concept and preliminary documentation
1.00	December 2006	First non-public release
1.01	January 2007	Expanded, Edited & Reviewed
1.02	February 2007	Expanded, Edited & Reviewed
1.03	March 2007	Expanded, Edited & Reviewed
1.04	April 2007	Expanded, Edited & Reviewed
1.05	April 2007	Expanded, Edited & Reviewed
1.06	June 2007	Expanded, Edited & Reviewed
1.07	June 2007	Expanded, Edited & Reviewed
1.08	July 2007	Expanded, Edited & Reviewed
1.09	August 2007	Expanded, Edited & Reviewed
1.10	September 2007	Expanded, Edited & Reviewed
2.00	October 2007	**Completely Edited and Reviewed** First public release (1st Edition) Printed and Published by "*Books on Demand*"
2.01	November 2007	Expanded, Edited & Reviewed
2.02	December 2007	**Edited and Reviewed** 2nd Edition Printed and Published by "*Books on Demand*"
2.03	September 2008	**Edited and Reviewed** Various corrections or clarifications, lots of additions, etc. 3rd Edition Printed and Published by "*Books on Demand*"
2009	September 2009	**Edited and Reviewed** Various corrections or clarifications, lots of additions, etc. 4th Edition Printed and Published by "*Books on Demand*"

STRYK System Improvement
Performance Optimization & Troubleshooting

Legal Notice

The information contained in this document represents the current view of *STRYK System Improvement* (SSI) on the issues discussed as of the date of publication and is subject to change at any time without notice.

This document and its contents are provided AS IS without warranty of any kind, and should not be interpreted as an offer or commitment on the part of SSI.

SSI cannot guarantee the accuracy of any information presented.

SSI MAKES NO WARRANTIES, EXPRESS OR IMPLIED, IN THIS DOCUMENT.

The descriptions of other companies' products in this document, if any, are provided only as a convenience to you. Any such references should not be considered an endorsement or support by SSI. SSI cannot guarantee their accuracy, and the products may change over time.

Also, the descriptions are intended as brief highlights to aid understanding, rather than as thorough coverage.

All trademarks are the property of their respective companies.

Microsoft®, Microsoft® Dynamics™ NAV, Microsoft® SQL Server™ 2000, Microsoft® SQL Server™ 2005 and Microsoft® SQL Server™ 2008 are registered trademarks of the Microsoft Corporation; all rights reserved.

The NAV/SQL Performance Field Guide
Version 2009

Table of Content

Imprint 2
Document Version 3
Legal Notice 4
Table of Content 5
About the Author 8
From the Author 9
Preface 10
Performance Toolbox 11
What is "Performance"? 12
   > Performance Troubleshooting 13
Finding "Bottlenecks" 15
   > Windows Performance Monitor 16
   > NAV Client Monitor & Code Coverage 23
   > SQL Profiler 24
      >> Basic Interpretation of Execution Plans 26
   > Correlation SQL Profiler and Performance Monitor 27
Avoiding the Trouble – Fundamental Setup 28
   > Server Hardware Environment 28
      >> Moving Databases 37
   > Client Hardware Environment 40
   > Citrix/Terminal Server Hardware Environment 41
   > Middle Tier Environment (NAV 2009) 42
   > Used Version and Edition of the Operating System (OS) 43
      >> Memory – 32bit Architecture 43
      >> Memory – 64bit Architecture 46
      >> Processors 47
   > Used Version and Edition of the SQL Server 49
   > Configuration of SQL Server and Database 51
      >> SQL Server Instance Settings 51
      >> Database Settings 54
      >> NAV Client Settings 57
      >> Example TSQL for Configuration 61
   > Optimizing "tempdb" 63
Fixing the Trouble – Erasing problems 64
   > Structure of Indexes in NAV 65
      >> Reduction of the number of Indexes 69
      >> Optimizing the order of fields 70
      >> Remove UNIQUE Flag and added PK Fields 75
      >> Define optimized Fill-Factors for the indexes 77

STRYK System Improvement
Performance Optimization & Troubleshooting

>> Index Statistics 83
>> Move indexes to a dedicated File-Group 85
>> Specific Hints for Index Creation in SQL Server 2005/2008 87
> Structure of SIFT Indexes in NAV 88
>> Moving SIFT to a dedicated File-Group 92
> Structure of VSIFT Views in NAV 94
>> Comparing SIFT and VSIFT 96
>> Replacing SIFT/VSIFT using Included Columns 99
> Design of C/AL Code 100
>> Querying SQL Server 101
>> Cursor Handling 104
>> Using SQL Server code 105
>> Linked Objects 108
>> Using Temporary Tables 112
> Miscellaneous Issues 114
>> The used C/SIDE Version 114
>> The DBCC PINTABLE feature 115
>> NAV "Table Optimizer" 116
>> Index Hinting 116
>> Different Query Performance in SQL 2000 and 2005 117
Getting rid of Locks, Blocks and Deadlocks 118
> Locking Mechanisms 118
>> Lock Escalation 118
>> Implicit Locking 119
>> Explicit Locking 119
> Using GUID 122
> Using AutoIncrement 124
> Forcing Row-Locking 125
> Setting Lock Granularity 126
> Block Detection 127
> Deadlocks 129
>> Index Optimization 130
>> Serialization / Semaphore 130
>> Common Locking Order 131
>> Transaction Speed 131
>> Investigating Deadlocks 131
Database Maintenance 135
> Maintenance Plan 135
> SQL Server Agent Jobs 137
> Backup Strategy 140
Parameter Sniffing 142
High Availability & Failover Strategies 146

- > Failover Clustering 146
- > Database Mirroring 147
- > Transaction Log Shipping 148
- > Database Snapshots 149
- > Replication 150

Miscellaneous 152
- > Ghost Cleanup 152
- > Named Pipes 152
- > Windows Registry & User Profiles 152
- > 32bit Applications on 64bit Servers 152
- > Dynamic Management Views 153

Additional Resources 154
Index 155
Appendix A – System Checklists 160
- Part A - SQL Server Configuration 160
- Part B – Database Configuration 165
- Part C – Database Maintenance 168
- Part D – Performance Monitor 169
- Part E – SQL Profiler 171

Appendix B – Version Lists 172
- MS Dynamics NAV (Navision) C/SIDE Versions 172
- MS Dynamics NAV (Navision) Database Versions 175
- MS SQL Server 2000 Versions 177
- MS SQL Server 2005 Versions 178
- MS SQL Server 2008 Versions 179

STRYK System Improvement
Performance Optimization & Troubleshooting

About the Author

Jörg A. Stryk, born 1971, works with *MS Dynamics NAV* – formerly *Navision* - since 1997, knowing the product since version 1.20.

Being *Project-Manager*, *Consultant*, *Developer* or *Supporter* on MS Partner-site and NAV Customer-site, as well, he has thoroughly explored nearly all areas from the application and its technology.

Since 2003 he is focusing on *Performance Optimization* and *Troubleshooting*, especially when it comes to *NAV* with *MS SQL Server*.

As *Microsoft Certified Professional (MCP)* and *Microsoft Certified Business Management Solution Specialist (MCBMSS)* he supports MS Dynamics Partners and Customers within their NAV projects – worldwide!

In recognition of his valuable participation in various NAV communities he received the Microsoft award *Most Valuable Professional (MVP)* for *MS Dynamics NAV* in 2007, 2008 and 2009.

The British Dynamics Community "*Dynamics World*" (www.dynamicsworld.co.uk) voted him on rank #38 of the "*Top 100 Most Influential People in the Dynamics World 2009*".

The NAV/SQL Performance Field Guide
Version 2009

From the Author

Dear Troubleshooter,

This "*Performance Field Guide*" is actually the attempt to put all essential information about NAV/SQL performance optimization - gathered over the years in my projects, after investigating hundreds of internet resources, bothering plenty of experts and finally scribbled down on countless papers - in one single document.

It is the "brain dump" I use when fixing problems, just as reminder and template – nobody could remember everything.
Whenever I learn something new, I add it to the PFG or change it, so it should include my most recent knowledge about problems and solutions.

The current version of the PFG is a solid basis for all those who want to – or have to – deal with performance optimization, but updated versions will follow, for sure.

All the "problems and solutions" are intensively discussed in various community forums on the Internet, so different people may have different opinions about certain issues. I encourage you to participate in these discussions!

I keep this booklet in English to provide the information to a wider audience; please have in mind that this is not my native language and I won't hire a professional translator - I apologize for bad grammar or weird vocabulary ;c)

Cheers,

Preface

Microsoft Dynamics NAV (formerly known as *Navision Financials*, or *Navision Attain* or *Microsoft Business Solutions Navision* etc.) is one of the leading ERP systems on the market. It was designed to run on its own "black-box" database, and it was running quite good. Well, since Microsoft took over, this "native" database is actually "*sentenced to death*" – we can be pretty sure it will vanish in future, it's just a matter of time.

Microsoft SQL Server is the favorite database for NAV, which is absolutely the right decision. It could be said, that SQL Server is the right database for NAV, but NAV is not completely the right application for SQL Server (yet).
To protect the investments of NAV customers, MS decided to create the application in a way to be compatible with both database servers – C/SIDE and SQL Server. Hence, currently NAV is a kind of "compromise-solution" which provides advantages – and disadvantages – for both servers.
So yes, there are issues with NAV and SQL Server.
Microsoft is aware of those "insufficiencies" and is proceeding to fix the problems, actually NAV is getting better with every Release, Service Pack or Update.
Unfortunately, the progress is somewhat too slow for those who have to struggle daily with the system, especially the ones running older versions of NAV. Hence, we have to fix the most crucial problems ourselves – Partners & Customers.

"*The NAV/SQL Performance Field Guide*" – short PFG – is a booklet which should explain in a simple and easily understandable way the known problems regarding NAV on SQL Server and further, it should give feasible advice, solutions and recommendations to fix and avoid problems.
The PFG is not a too technical documentation and is not a replacement for existing documentation on NAV and SQL Server.
This also means that some "background knowledge" is presumed; when using the PFG you have to know about NAV, C/SIDE & C/AL, SQL Server, TSQL, Windows OS, Hardware, etc..

Everything you use or implement from the PFG you do at your own risk. No guarantee or warranty, errors and omissions excepted. No support.
If you are not sure if/what to implement - if you have any doubts - then don't do it! Ask for support, else you might worsen existing problems or raise new ones!

Get help at http://www.stryk.info/

The NAV/SQL Performance Field Guide
Version 2009

Performance Toolbox

Many of the described improvements only could be implemented by using the right *utilities*!

STRYK System Improvement provides a collection of very useful tools to optimize the system's performance:

"*The NAV/SQL Performance Toolbox*"

Whenever you see in the "*NAV/SQL Performance Field Guide*" this picture …

… there is an adequate tool in the **Performance Toolbox**!

The "*NAV/SQL Performance Toolbox*" is not a product which could be directly purchased – it is *exclusively* distributed by competent and authorized MS Dynamics partners!

Learn more about PTB at http://www.stryk.info/english/toolbox.html

STRYK System Improvement
Performance Optimization & Troubleshooting

What is "Performance"?

Yeah, that's a tricky and mostly a philosophical question ... not easy to answer. In our matters "performance" is actually defined with two essential criteria:

Throughput and Response-Time

Throughput = The **volume** of data which is processed within a period.
Response-Time = The **duration** for processing a defined amount of data.

So, regarding **performance optimization** we could achieve two goals:

Either

Process a defined volume of data within a <u>minimum period</u> of time

Or

Process a <u>maximum volume</u> of data within a defined period of time

To optimize the performance for transactions or processes it has to be <u>decided</u> **individually** which principle to follow. It is <u>not</u> possible to "*process a maximum volume of data in a minimum period of time*" – that's non-sense (but unfortunately that's what many users are asking for).

Simplified:

> **Performance = Throughput / Response-Time**

Hence, Performance will increase either by increasing the Throughput or decreasing the Response-Time!

The trick with performance troubleshooting is to find the right **balance** of both principles, which requires a lot of experience and know-how!

That's the challenge!

> Performance Troubleshooting

Performance Troubleshooting is complex and time consuming; it could be considered a mixture of "**Art & Science**". This booklet just could give some information for the "*Science*" part (compressed & simplified, not the whole truth), but not for the "*Art*" – the experience, knowledge and feeling about "*when to implement what*". The risk is, if "*prescribing the wrong medicine*" this would worsen existing problems and raise new trouble.
So it's a serious advice: If you have any doubts, then don't do it! Ask for assistance!

Proceeding troubleshooting:

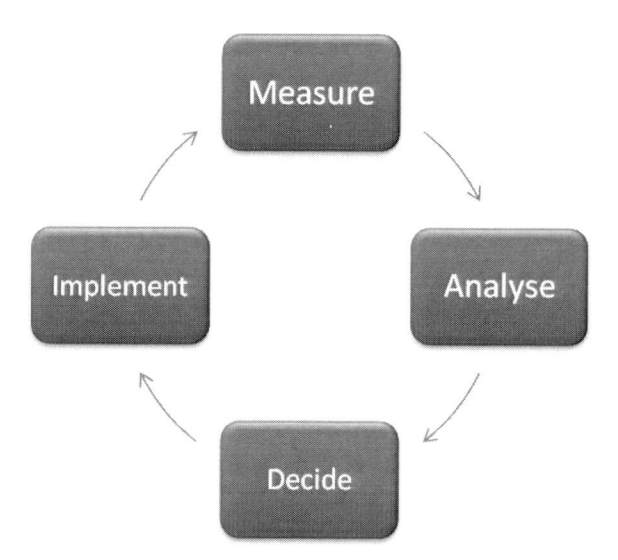

First **measure** the "facts": *Response Times* and *Data-Volume*. **Analyze** the facts to determine what could be the problem. Investigate the available options to solve the problem and **decide** what to implement and when. Once a potential solution was **implemented** verify the result by measuring again …

STRYK System Improvement
Performance Optimization & Troubleshooting

Areas to investigate and improve:

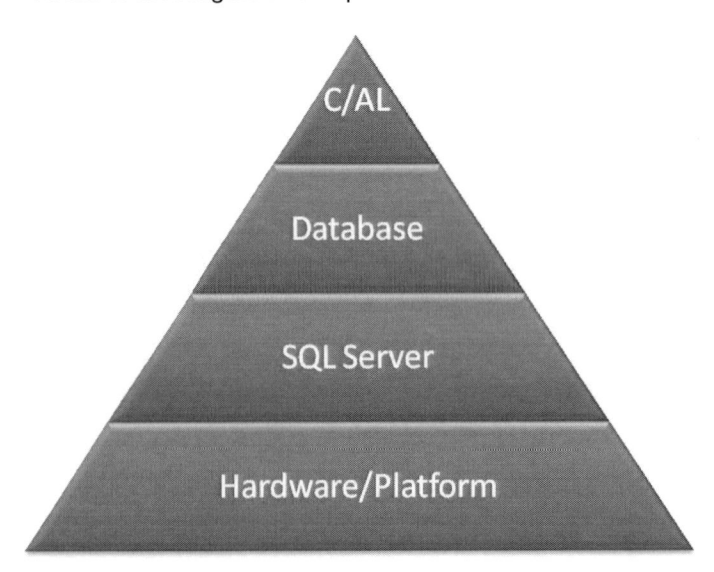

For a successful optimization it is necessary to start from the bottom – the platform.
It is pointless to "raise a house on a weak fundament"; or:

You don't have to fiddle with the C/AL Code if actually there is just not enough RAM. In this case installing more RAM will give a higher benefit than redesigning some code.

Regarding the **time/effort** which has to be spent to fix things it's usually quite the opposite way.

Fixing the platform – e.g. adding RAM - is mostly faster and easier than changing C/AL – e.g. reprogramming posting routines.

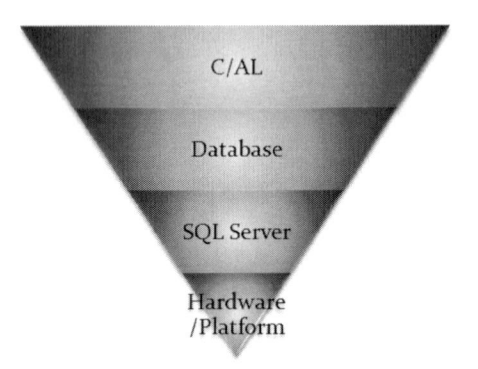

Finding "Bottlenecks"

Usually the first one who is confronted with "performance issues" is … the user: the one who daily works with the system, the one who should have the deepest insight into his business. Whenever performance is changing – decreasing – he will notice first and complain "*things are so slow*" (The funny thing is, that users basically never mention if things are getting faster than usual). Even though this is a quite subjective point of view, real life has proved that this kind of "performance monitoring" is somewhat reliable and worth to be investigated.
Now it's up to change thus subjective view into some objective facts, therefore some **tools** are feasible:

- Windows Performance Monitor
- NAV Client Monitor & Code Coverage
- SQL Profiler

STRYK System Improvement
Performance Optimization & Troubleshooting

> Windows Performance Monitor

The **Windows Performance Monitor** (`perfmon.exe`) is a very convenient utility to measure and report a variety of system- and application-indicators. The following describes some of the <u>most important</u> **indicators**:

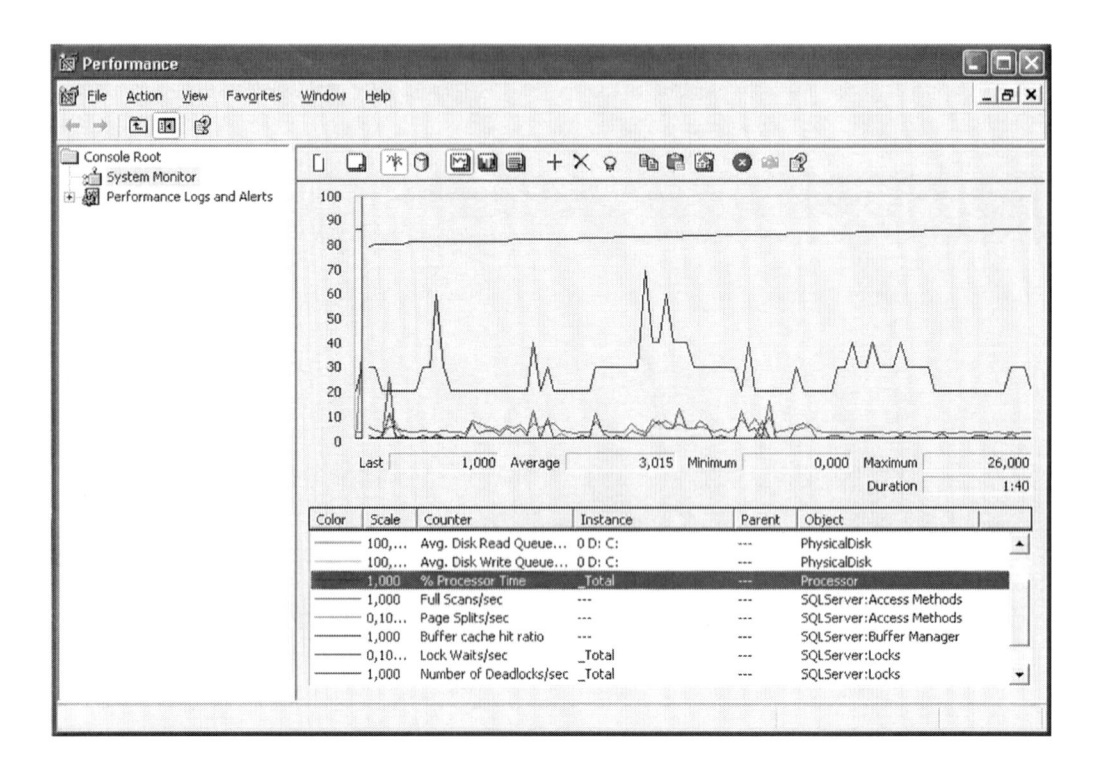

The NAV/SQL Performance Field Guide
Version 2009

Basic Indicators

Object	Counter	Instance	Best Value	Explanation
Memory	Available MBytes	n/a	> 10	RAM/Memory is the most important component for the SQL Server. The more – the better! When running out of RAM the SQL Server has to read data from disk which decreases performance dramatically. **Add RAM!**
Memory	Pages/sec	n/a	< 25	Number of pages transferred from/to disk to resolve hard page faults. The higher this number grows, the more disk I/O occurs which reduces performance. **Add RAM!**
Physical Disk	Avg. Disk Read Queue Length	NAV Database files NAV Transaction Log files	< 2 per disk	The average number of read requests that were queued for the selected disk during the sample interval. The longer the queue length, the longer SQL Server has to wait for data from the disks. **Change disk sub-system!**
Physical Disk	Avg. Disk Write Queue Length	NAV Database files NAV Transaction Log files	< 2 per disk	The average number of write requests that were queued for the selected disk during the sample interval. The longer the queue length, the longer SQL Server has to wait for data from the disks. **Change disk sub-system!**
Physical Disk	Avg. Disk Sec/Transfer	NAV Database files NAV Transaction Log files	< 0,015	The average duration of disk-transfers (per disk). **Change disk sub-system!**
Physical Disk	Transfer/Sec	NAV Database files NAV Transaction Log files	< 120	The rate of read- and write-transfers per disk. **Change disk sub-system!**
Physical Disk	Time %	NAV Database files NAV Transaction Log files	< 50	Time spent to read or write data from/to disk. See "Processor: % Privileged Time" **Change disk-subsystem!**

STRYK System Improvement
Performance Optimization & Troubleshooting

Object	Counter	Instance	Best Value	Explanation
Processor	% Processor Time	Each CPU	(< 80) < 30	The percentage of elapsed time that the processor spends to execute a non-Idle thread. Ideally the Processor Time is at 20% to 30% at average, so the system has enough reserves to handle peak loads fast. Latest, if the avg. is at 80% you have to **add more CPU!**
Processor	% Privileged Time	Each CPU	< 10	Processor Time (%) spent on CPU Kernel. If value is greater than 20 and "*Disk Time %*" is greater 55 there is an I/O problem. **Change disk sub-system!**
System	Processor Queue Length	n/a	< 2	The number of threads in the processor queue, to be divided by the number of processors servicing the workload. The higher this number, the more threads are waiting and delayed. **Add more CPU!**
System	Context Switches/sec	n/a	< 8000 per CPU	The combined rate at which all processors on the computer are switched from one thread to another. Switching threads is very costly and has to be avoided. The amount of Context Switches could be reduced by setting the **Affinity Mask** or when enabling **NT Fibers** (not possible with NAV & Windows Logins)
Network Interface	Output Queue Length	NAV Client-Server Connection	< 2	The length of the output packet queue (in packets). The higher this number, the more packets are waiting/delayed. **Change network interface, increase bandwidth.**
SQL Server Access Methods	Full Scans/sec	NAV database	0	This should never happen, because this increases the transaction's duration and causes blocks. **Optimize Index structure and C/AL code.**

The NAV/SQL Performance Field Guide
Version 2009

Object	Counter	Instance	Best Value	Explanation
SQL Server Access Methods	Page Splits/sec	NAV database	0	Number of page splits occurring as the result of index pages overflowing. Reallocating pages is time consuming. **Defrag indexes and optimize Fill-Factors.**
SQL Server Buffer Manager	Free Pages	n/a	> 640	Number of pages currently not used by the system. **Add RAM!**
SQL Server Buffer Manager	Buffer Cache Hit Ratio	n/a	> 90	Percentage of pages that were found in the buffer pool without having to incur a read from disk. If this rate drops, the disk I/O will increase – slowing down the system. **Add more RAM!**
SQL Server Buffer Manager	Page Life Expectancy	n/a	> 300	Time a page remains in the cache until it is overwritten. Regard *"Memory Grants Pending"*. **Add more RAM!**
SQL Server Locks	Lock Request/sec	Total	0	Number of new locks and lock conversions requested from the lock manager. Depends on application. **Optimize C/AL Code.**
SQL Server Locks	Lock Waits/sec	Total	0	Number of lock requests that could not be satisfied immediately and required the caller to wait before being granted the lock. Depends on application. **Optimize C/AL Code and Index structure.**
SQL Server Locks	Number of Deadlocks/sec	Total	0	Number of lock requests that resulted in a deadlock. Depends on application. **Optimize C/AL Code and Index structure.**
SQL Server Memory Manager	Target Server Memory (KB)	n/a	= physical RAM – 1GB	Total amount of dynamic memory the server is willing to consume. **Check configuration!**
SQL Server Memory Manager	Total Server Memory (KB)	n/a	= Target Server Memory	Total amount of dynamic memory the server is currently consuming. Should equal the Target Server Memory. **Check configuration!**
SQL Server Memory Manager	Memory Grants Pending	n/a	0	Number of processes waiting for memory grants. **Add more RAM!**

Optional Indicators

Other counter worth for inspection

Object	Counter	Instance	Explanation
SQL Server Databases	Data File(s) Size (KB)	NAV database	The cumulative size of all the data files in the database.
SQL Server Databases	Log file(s) Size (KB)	NAV database	The cumulative size of all the log files in the database.

SQL Server Databases	Log Growths	NAV database	Total number of log growths for this database. Too many growth decrease performance.
			
			Check the number of "**Virtual Logs**" (VLogs) by executing this TSQL:
			
			`use [Navision]  -- change db name`
			`go`
			`dbcc loginfo`
			
			The number of lines shown represents the number of VLogs.
			(See also below about "Auto Growth")

SQL Server Databases	Percent Log Used	NAV database	The percent of space in the log that is in use.
SQL Server Databases	Transactions/sec	NAV database	Number of transactions started for the database.
SQL Server General Statistics	User Connections		Number of users connected to the system.

When monitoring the system with the *PerfMon* it is necessary to get the "*whole picture*"; means the monitoring has to be performed at different times with different loads for a longer, **representative** period to get **reliable figures**. Also, all indicators have to be analyzed **together** as many of them are related; concentrating on just a single issue could be misleading.

PerfMon could be scheduled and be used to generate automatic reports. Be careful, the monitoring itself is slightly decreasing performance, so it should not be a permanent measurement.

The NAV/SQL Performance Field Guide
Version 2009

To get a quick overview it is recommended to switch to "**Report**" mode, there showing the "*Current*" or "*Average*" figures:

```
Memory
    Available MBytes                    2784
    Pages/sec                          0,000

MSSQL              :Access Methods
    Full Scans/sec                     1,000
    Page Splits/sec                    0,000

MSSQL              :Buffer Manager
    Buffer cache hit ratio            99,739
    Free pages                          1695
    Page life expectancy                7999
    Readahead pages/sec                0,000

MSSQL              :General Statistics
    Processes blocked                      0
    User Connections                      64

MSSQL$             Locks              _Total
    Lock Requests/sec                188,038
    Lock Wait Time (ms)                0,000

MSSQL$             :Memory Manager
    Memory Grants Pending                  0
    Target Server Memory (KB)       21099392
    Total Server Memory (KB)        21099392

MSSQL$             :Plan Cache        _Total
    Cache Hit Ratio                   95,136

Network Interface         Broadcom NetXtreme Gigabit Ethernet   Broadcom NetXtreme Gigabit Fiber
    Current Bandwidth                       1000000000                          1000000000
    Output Queue Length                              0                                   0

Paging File               \??\G:\pagefile.sys
    % Usage                            17,201

PhysicalDisk              0 E:          1 G:        2 H:        3 F:        4 C: D:
    Avg. Disk Read Queue Length      0,005         0,000       0,056       0,000       0,000
    Avg. Disk sec/Transfer           0,005         0,000       0,009       0,000       0,005
    Avg. Disk Write Queue Length     0,000         0,000       0,000       0,000       0,009

Processor                 _Total           0           1           2           3
    % Privileged Time     3,906         3,125       7,812       1,562       3,125
    % Processor Time      7,423        10,939      10,939       1,564       6,251

System
    Context Switches/sec         12398,508
    Processor Queue Length               0
```

Performance Monitor – "Graph" Mode

When viewing a "*Counter Log*" as "**Graph**" in perfmon it might look very annoying, displaying lots of vertical lines.
This is actually a bug in Windows Server 2003 which could be fixes by adding this to the Windows Registry:

```
[HKEY_CURRENT_USER\Software\Microsoft\SystemMonitor]
"DisplaySingleLogSampleValue"=dword:00000001
```

See also: http://support.microsoft.com/kb/283110

Reloading Performance Libraries

Once in a while it could happen that the performance indicators of the SQL Server are not available/visible. This could be caused by missing Registry Entries.
To fix this, the relevant performance libraries could be "reloaded" using the **lodctr.exe** feature:

```
lodctr.exe /r:sqlctr.ini
```

Performance Counter – Dynamic Management View (DMV)

With SQL Server 2005/2008 it is also possible to view the SQL Server performance counters e.g. in "Management Studio":

```
SELECT * FROM sys.dm_os_performance_counters
```

> NAV Client Monitor & Code Coverage

When speaking about "**Client Monitor**" It's actually one name for two applications. The **basic** CM is part of the NAV client (Tools\Client Monitor). With the "*Performance Troubleshooting Tools*" came some **enhanced** features which are based on this CM but providing far better analyzing options.
Usage and purpose of these Client Monitor features are thoroughly described in the "Performance Troubleshooting Guide" (`w1w1perf.pdf`) shipped with the tools; thus there's no further explanation here.

The advantage of using the **Client Monitor** incl. **Code Coverage** is, that here it could be determined clearly where in the C/AL source code a problem occurs/is raised.

Unfortunately there are some limitations of the CM:

- When monitoring big transactions, the NAV client – incl. CM – will crash
- CM monitoring cannot be performed automatically
- CM cannot "snap in" a running transaction
- CM cannot monitor Non-GUI processes, e.g. running on a NAV Application Server (NAS)

Due to these disadvantages it is feasible to use the **SQL Server Profiler** instead or in addition.

Once you have identified a problem in *SQL Profiler* and related it to the "responsible" NAV code, then *Client Monitor/Code Coverage* might be feasible to further investigate the problem.

> SQL Profiler

The SQL Profiler provides all required information for performance troubleshooting and much more. Actually all the disadvantages of the Client Monitor are eliminated: one could trace anything, anytime – online or unattended/scheduled. Read the SQL Profiler manuals and SQL Server "Books Online" for details. The disadvantage if the Profiler is actually the advantage of the Client Monitor: it's hardly possible to relate the profile record to the right piece of C/AL code.
So together – Client Monitor and SQL Profiler – will give the best results to detect problems within the application.

Recommended Setup of the SQL Profiler:

Events	Columns	Filters

Stored Procedures
  RPC: Completed
  SP: Completed
  SP: StmtCompleted
 	
TSQL
  SQL: BatchCompleted
  SQL: StmtCompleted
	SPID
	EventClass
	TextData
	Reads
	Writes
	CPU
	Duration
	Start Time
	End Time
	LoginName
	HostName
	Application Name
	Database Name
		Application Name
		  Not Like %SQL%
		
		Reads
		  Greater than or equal
		  <Threshold>
		
		Duration
		  Greater than or equal
		  <Threshold>

Optional / On Demand		

Performance
  Showplan Text Unen-
  coded
  (SQL 2000)
 	
  or
 	
  Showplan XML
  (SQL 2005/2008)		

The NAV/SQL Performance Field Guide
Version 2009

The problems that could be detected with the Profiler are (among others)

➢ **High Number of Reads**
This is the most important issue to fix! It usually indicates that "wrong" – suboptimal – indexes were used and that the Execution Plan is causing trouble, e.g. performing *Scans* instead of *Seeks*.
Potential Problems: Reads >= 1000

➢ **High Number of Writes**
This could indicate that the "**Costs per Record**" are too expensive; means that too many indexes are maintained ("over-indexed") or that too many SIFT buckets are filled.
Potential Problems: Writes >= 50

➢ **Long Duration**
Usually according to Read/Write problems, but could also be caused by blocks
Potential Problems: Duration >= 50

➢ **"Bad" Execution Plans**
Usually related to Read problems. *Scans* could also be responsible for blocks.
(Further information see below)

Experienced Profiler users can also learn **a lot more** from this utility; for example the handling of cursors gives a hint about "bad" programming in NAV, etc..

As the Profiler traces can be saved – preferably to file – the monitoring could be performed unattended and troubleshooting could be performed "asynchronous" at a different time or system.
The TRC files could be loaded to a table afterwards for better investigation:

```
SELECT * INTO MyTraceTable
FROM ::fn_trace_gettable('<Path>\<Filename>.trc', default)
```

(**Caution:** The TRC files could also be fed to the "**Tuning Advisor**", but unfortunately its proposals are somewhat critical and should only be implemented if an expert has approved them)

According to the "**unattended use**" of the Profiler the feature of exporting the Trace Setup into a **TSQL** script has to be mentioned. Thus, a Trace could be configured in the Profiler, and then exported to TSQL which could be executed via a SQL Server Agent Job!

STRYK System Improvement
Performance Optimization & Troubleshooting

>> Basic Interpretation of Execution Plans

Operation	Comment
Bookmark Lookup	After reading a Non-Clustered Index the records are looked up from the Clustered-Index.
Clustered Index Scan	A Clustered Index – or parts of it – is scanned. This is time- and resource- consuming and could run into blocking situations. **Optimize Index and C/AL code.**
Clustered Index Seek	Data is retrieved from Clustered Index. Usually best performance.
Filter	Data is filtered on the WHERE clause. Could indicate that a sub-optimal Index is used for retrieval. **[Optimize C/AL code.]**
Index Scan	A Non-Clustered Index – or parts of it – is scanned. This is time- and resource- consuming and could run into blocking situations. **Optimize Index.**
Index Seek	Data is searched via Non-Clustered Index. If the number of Reads is too high, then the chosen Index is sub-optimal. **[Optimize Index.]**
Sort	(Re-)Sorting operation according to the ORDER BY clause. If costs are too high, the "gap" between used Index and sorting could be too big. **[Optimize Index and C/AL code.]**
Table Scan	The complete table is scanned. **Worst performance!** This is time- and resource- consuming and could run into blocking situations. **Optimize Index and C/AL code.**

Here it is crucial to interpret the **combination** of all operations; and also the query itself!

The NAV/SQL Performance Field Guide
Version 2009

> Correlation SQL Profiler and Performance Monitor
(SQL Server 2005/2008)

With SQL Server 2005/2008 it is possible to correlate the trace files of the SQL Profiler and the Windows Performance Monitor.
This allows a more detailed analysis; investigating the platform's performance in <u>direct</u> relation to the relevant SQL query.
This e.g. could show how huge numbers OF Read will affect the *"Buffer Cache Hit Ratio"* and/or disk I/O.

1. Setup automatic PerfMon **Counter Log**, writing its output into file
2. Setup SQL Profiler **trace**, writing its output into file
3. Run the processes to be investigated
4. Stop PerfMon and SQL Profiler Traces
5. Open Profiler trace-file with SQL Profiler
6. Select *"File - Import Performance-Data"* and pick the log-file from PerfMon
7. Choose performance counters to display

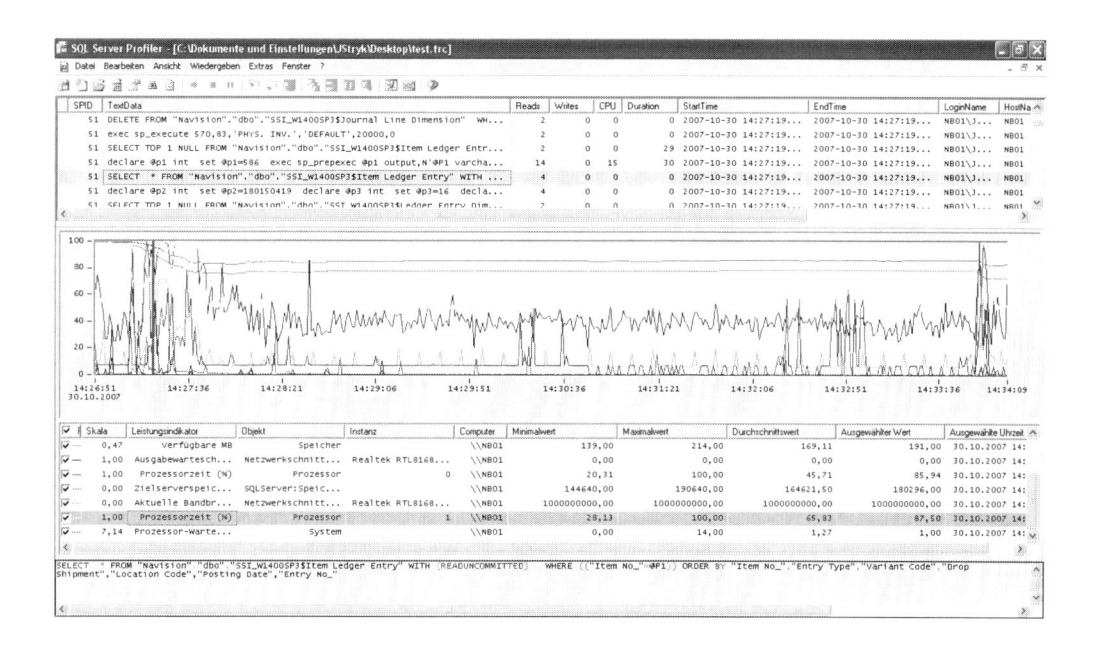

Avoiding the Trouble – Fundamental Setup

Before fixing any kind of specific problems it is crucial to have a sufficient system environment; means the fundamental basis of the system has to be set up and configured properly.

These "fundamentals" are basically

- Server Hardware Environment
- Client Hardware Environment
- Citrix/Terminal Server Hardware Environment
- Middle Tier Environment (NAV 2009)
- Used Version and Edition of the Operating System (OS)
- Used Version and Edition of the SQL Server
- Configuration of SQL Server and Database

> Server Hardware Environment

Microsoft provides the "**Hardware Sizing Guide**" for NAV giving good advice about which hardware to use. The problem with these advices is (and it's the same issue with lots of "official" performance statistics), that they refer only to the **number of concurrent users**, e.g. when deciding how many CPU to implement, etc.. Unfortunately, this is just "half of the truth". The number of concurrent users is only usable if the **transaction volume** is in direct proportion to it; means for example *1 user creates 1 transaction, 2 users create 2 transactions*, etc.. So, it's not the number of users to take into account, it is the **expected transaction volume**!

For example, there could be companies with just 10 users, but generating 10.000 order lines per day, while on the other hand there might be companies with 200 users handling 100 lines per day.

So "Users" should be specified as "*Light*", "*Medium*" or "*Heavy*".

When sizing a NAV & SQL environment it is absolutely crucial to **investigate thoroughly** this expected transaction volume, the number of documents created and posted, the length of the documents etc.!

The NAV/SQL Performance Field Guide
Version 2009

Absolute **minimum** requirements (*DB size < 50GB, low transaction volume*):

Component	Recommendation	Comment
Architecture	32bit	Better use 64bit
CPU	1 x QuadCore	
RAM	4 GB (better 8 GB)	The more, the better! Regard OS limits!
Network	Gigabit Ethernet	In some cases 100Mbit might be OK
Operating System	Windows Server 2003 Standard	Regard limitations regarding RAM or CPU usage
Disk Subsystem	C:\ (RAID1 [2 x HDD])	Operating System, Swap-File, SQL Server Program Databases **master**, **model**, **msdb** and **tempdb** (mdf & ldf)
	D:\ (RAID1 or 10 [2-6 x HDD])	**NAV** Database (mdf/ndf)
	E:\ (RAID1 [2 x HDD])	**NAV** Database (ldf)

The following should give an **example** about a feasible "mid-size" system (*DB size > 50GB & < 100GB, medium transaction volume*) environment (as hardware improves daily, this could be outdated as it was written down):

Component	Recommendation	Comment
Architecture	64bit	
CPU	1 or 2 x QuadCore	1 physical CPU per 100 users
RAM	8 GB	The more, the better!
Network	Gigabit Ethernet	
Operating System	Windows Server 2003 Standard 64bit	Regard limitations regarding RAM or CPU usage
Disk Subsystem	C:\ (RAID1 [2 x HDD])	Operating System, Swap-File, SQL Server Program
	D:\ (RAID1 [2 x HDD])	Databases **master**, **model** and **msdb** (mdf & ldf)
	E:\ (RAID1 [2 x HDD])	Database **tempdb** (mdf & ldf)
	F:\ (RAID 10 [4-8 x HDD])	**NAV** Database (mdf/ndf)
	G:\ (RAID1 or 10 [2-6 x HDD])	**NAV** Database (ldf)

STRYK System Improvement
Performance Optimization & Troubleshooting

The following should give an **example** about a feasible "large-size" system (*DB size > 100GB, high transaction volume*) environment (as hardware improves daily, this could be outdated as it was written down):

Component	Recommendation	Comment
Architecture	64bit	
CPU	2 or 4 x Quad Core	1 physical CPU per 100 users
RAM	16 to 32 GB	The more, the better!
Network	2 Gigabit Ethernet	
Operating System	Windows Server 2003 Enterprise 64bit	Regard limitations regarding RAM or CPU usage
Disk Subsystem	C:\ (RAID1 [2 x HDD])	Operating System, Swap-File, SQL Server Program
	D:\ (RAID1 [2 x HDD])	Databases **master**, **model** and **msdb** (mdf & ldf)
	E:\ (RAID1 [2 x HDD])	Database **tempdb** (mdf & ldf)
	F:\ (RAID10 [4-10 x HDD])	**NAV** Database (mdf/ndf)
	G:\ (RAID10 [4-8 x HDD])	**NAV** Database (ldf)

The most critical part is the disk subsystem. SQL Server processes all transactions in its cache, so the more cache – RAM – is available the higher the performance. The disks are the slowest part as they have to do mechanical work. Hence, an insufficient disk environment will cause delays for the whole system!

Dynamics Partners could also contact the "**Technical Presales Advisory Group**" (TPAG) to ask about proposals for appropriate sizing especially with large NAV installations.

See: https://partner.microsoft.com/global/productssolutions/40023009

Virtual Environments

This is an intensively discussed issue, there are Pro's and Con's, of course – I don't want to discuss this here.
One key-fact is this: with virtualization more hardware resources are required than with a physical environment, else performance will be degraded.

I recommend to use a real physical environment for "productive" Database Servers.

The NAV/SQL Performance Field Guide
Version 2009

Using RAID (*Redundant Array of Independent Disks*) is strongly recommended to store the data safe and fast:

RAID 0: *Striping*. The data is distributed over multiple spindles. Thus, the mechanical work is split, the transaction is performed faster. It is better to have many smaller disks than fewer larger disks.

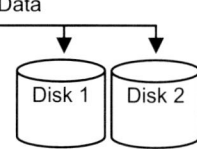

RAID 1: *Mirroring*. The data is copied – mirrored – from one spindle to another. If one disk fails, the mirror could take over, no data is lost.

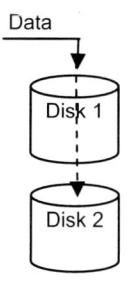

RAID 10: *Striping & Mirroring*. The combination of RAID 0 and RAID 1. Here data is stored safe because it's mirrored and fast as it is distributed over multiple spindles.

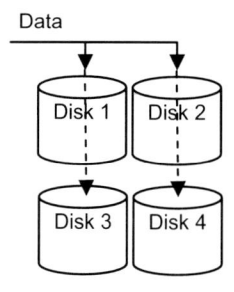

In some older documentation also **RAID 5** (*striping with parity*) is mentioned as an alternative option, but experience has shown that this is definitely not recommendable as it performs too slow.

STRYK System Improvement
Performance Optimization & Troubleshooting

Some brief explanations about the database distribution:

Pubs and **Northwind** are just demo-databases. Actually they could be deleted (or not installed at all), but at least they should be placed where they don't do any harm.

On large/heavily used systems it could be feasible to place the system databases - e.g. **master** (THE database, stores all system information, database catalogues, logins, etc.), **model** (the template for all databases) and **msdb** (the database of the SQL Server Agent, includes Jobs, Operators, DTS/SSIS, etc.) - on dedicated disks (RAID1).
The hidden system database **mssqlsystemresource** should always be placed in the same folder as **master**!

The **tempdb** is crucial regarding performance. It is (re-)created whenever the SQL Server starts up. It is used to store temporary result sets, etc.. While it is used moderately with "normal" NAV business, it could be heavily used when dealing with large result-sets and sort operations. A slow *tempdb* db cause delays for the whole system. RAID1 preferred, but also RAID0 could be an option.

The **Database** files of the **NAV** db have to be placed on a dedicated disk. RAID 1 to keep the data safe; even better RAID 10 to have it fast and safe.
The **Transaction Log** files of the **NAV** must be placed on a dedicated disk. All transactions are depending on the physical read/write processes of the TLog. Further, the TLog is required to restore the database up to the very last committed transaction. Hence, it must be the **fastest and safest** disk in the system! RAID 1 or RAID10.

Running a sufficient SQL Server environment is relatively costly regarding disk requirements. But any compromise here would have negative impact on the performance.

The NAV/SQL Performance Field Guide
Version 2009

Other considerations

The "make and model" of the disks is also very important:

SATA (*Serial Advanced Technology Attachment*) disks have a **MTBF** (*Mean Time Between Failure*) 10 times worse than **SCSI** (*Small Computer System Interface*). SATA disks will fail 10 times more often than SCSI disks.
Hence, it is strongly recommended to use **SCSI** disks for a database server!

Regarding SCSI technology **SAS** (*Serial Attached SCSI*, 3GBit/s) is preferable over U320 (Ultra 320 SCSI, 320 MB/s). If ever possible also opt for 15000 rpm over 10000 rpm disks.

The „spindle count" applies to advanced storage solutions like **SAN** (*Storage Area Network*) also. Do not expect your ERP solution on SQL to perform well only because you own a SAN!

If all ERP data is stored in a single SAN volume comprising only a few large disks, and other applications compete with ERP for disk I/O and bandwidth, ERP performance will suffer badly!

For an optimal ERP SAN remodel the database distribution schema outlined above on **dedicated physical disks** in your SAN and spread your ERP volumes over as many physical spindles as financially and technically possible.

A **NAS** (*Network Attached Storage*) should not be used to store databases as it performs too slow.

[Acknowledgement to Christian Krause, amball business software/Germany for reviewing this paragraph and adding his most appreciated insights!]

Solid State Disks

The slowest part of a SQL Server is the "*Disk-Subsystem*" because here the spindles have to perform mechanical work (turning spindle, moving read/write head), which is simply time consuming. **Solid State Disks** (SSD) do not have any moving mechanical parts, thus they could perform much faster than usual HDD!

But at this point in time SSD seem not to be ready to deal with high transaction volumes and large databases properly.
Also, at this point in time no long-term practice experience is available.

The costs for SSD are somewhat higher than for common HDD (but getting cheaper day by day), also due to the issue that SSD have limited **write-cycle capability**, thus the disks have to be replaced after some months.

Hence, SSD is coming up and it promises great improvements, but currently common HDD might be preferred.

The NAV/SQL Performance Field Guide
Version 2009

More about SAN (*Storage Area Network*)

Basically SAN are designed like this: an **array** of spindles is combined into one or more **aggregates/targets** (terminology depending on manufacturer). Thus, the "aggregate" is the physical layer of the SAN. Now such an aggregate could be divided into several logical **volumes** and **LUN** (*Logical Unit Number*).

These LUN are assigned to e.g. the SQL Server or other servers etc., e.g. Exchange, Sharepoint or used as Fileserver etc. – Example:

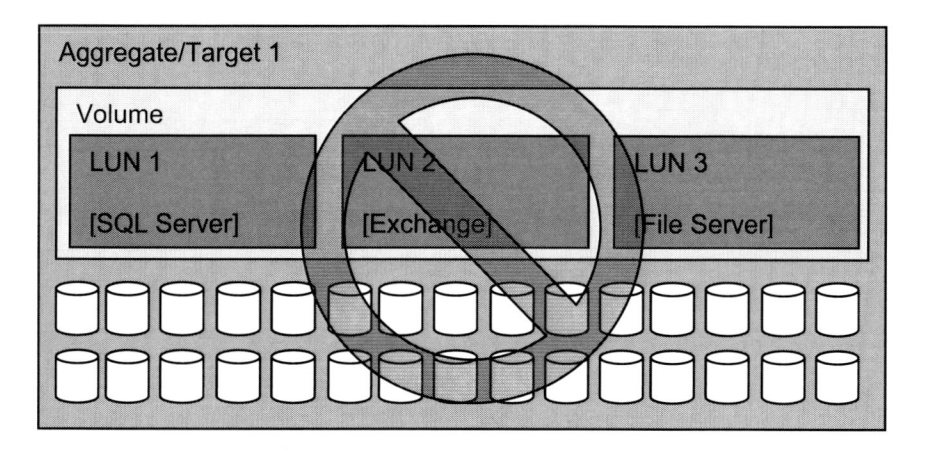

```
Aggregate/Target 1
  Volume
    LUN 1          LUN 2           LUN 3
    [SQL Server]   [Exchange]      [File Server]
```

This means that *one* aggregate (physical!) could be used by *multiple* applications, e.g. by SQL Server AND Exchange.

Hence, from a physical perspective the data from all LUN is *mixed* because the data is transferred from the SAN's cache to disk by FIFO (*First in First Out*) principle!

This only performs well up to a certain amount of data-/transaction-volume. If the SAN is heavily used it could be a problem when reading the data from the disks/spindles as it could be spread over wide areas within the aggregate – simplified: the SAN needs time to gather e.g. the SQL data together.

STRYK System Improvement
Performance Optimization & Troubleshooting

In this case it is necessary to have at least one **dedicated** aggregate for the SQL Server (high transaction volume), or even better (very high transaction volume) two aggregates: one for the databases, and one for the NAV Transaction Log.

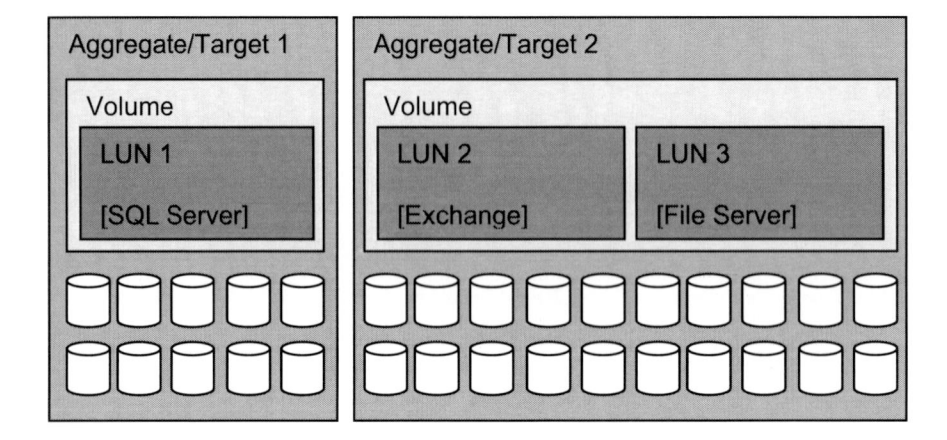

Aggregate/Target 1	Aggregate/Target 2	
Volume	Volume	
LUN 1	LUN 2	LUN 3
[SQL Server]	[Exchange]	[File Server]

>> Moving Databases

Once the SQL Server has been installed with default settings, - storing the databases in default locations -, the databases could be moved to another location, afterwards.

Before doing this, it is highly recommended to make **backups** of <u>all</u> databases, especially the "master" database!

For further information refer to the MS Support Knowledge Base:

http://support.microsoft.com/kb/224071

Verification of Database-Files:

```
USE [<Database>]
GO

EXEC sp_helpfile
GO
```

Moving "master":

1. Within the "*Configuration Manager*" change the Startup-Parameters of the SQL Server

    ```
    -d<NewPath>\master.mdf
    -e<NewPath>\ErrorLog
    -l<NewPath>\mastlog.ldf
    ```

2. Stop the SQL Server service
3. Move the files `master.mdf` and `maslog.ldf` to the new location
4. Start the SQL Server service

With SQL 2005/2008, once the "master" database was moved you should also move the files `mssqlsystemresource.mdf` and `mssqlsystemresource.ldf`, these belong to the (hidden) system-database "**MsSqlSystemResource**" (see "Books Online").

STRYK System Improvement
Performance Optimization & Troubleshooting

Moving "model":

1. Alter database

    ```
    USE master
    GO

    ALTER DATABASE model MODIFY FILE
    (NAME = modeldev, FILENAME = '<NewPath>\model.mdf')
    GO

    ALTER DATABASE model MODIFY FILE
    (NAME = modellog, FILENAME = '<NewPath>\modellog.ldf')
    GO
    ```

2. Stop SQL Server service
3. Move database files to new location
4. Start SQL Server service

Moving "msdb":

1. Alter database

    ```
    USE master
    GO

    ALTER DATABASE msdb MODIFY FILE
    (NAME = MSDBData, FILENAME = '<NewPath>\MSDBData.mdf')
    GO

    ALTER DATABASE msdb MODIFY FILE
    (NAME = MSDBLog, FILENAME = '<NewPath>\MSDBLog.ldf')
    GO
    ```

2. Stop SQL Server service
3. Move database files to new location
4. Start SQL Server service

Moving "tempdb":

1. Alter database

```
USE master
GO

ALTER DATABASE tempdb MODIFY FILE
(NAME = tempdev, FILENAME = '<NewPath>\tempdb.mdf')
GO

ALTER DATABASE tempdb MODIFY FILE
(NAME = templog, FILENAME = '<NewPath>\templog.ldf')
GO
```

2. Restart SQL Server service

Moving User Databases:

1. Detach Database

```
USE master
GO

EXEC sp_detach_db '<Database>'
GO
```

2. Move database to new location
3. Attach Database

```
USE master
GO

EXEC sp_attach_db '<Database>'
,'<NewPath>\<Database>.mdf'
,'<NewPath>\<Database>.ldf'
GO
```

> Client Hardware Environment

Again the "*NAV Hardware Sizing Guide*" gives some recommendations regarding Hardware for NAV client PCs; which is basically equal to the Operating System requirements (*"when it's OK for Windows, it should be OK for NAV, too"*). Unfortunately, this just refers to the **installation** of a NAV client and completely ignores the fact that the NAV client also has to **process** data, as well!

When running NAV, all business logic is handled on the *C/SIDE client*. Therefore all queried data needs to be transferred from the Server to this client – hence, the client PC <u>must</u> have the capacity to handle the **volume** of data!

If the PC has insufficient **Network**-connection, **CPU** power and **RAM** the performance could be dramatically decreased. The slower one Client could process, the higher could be the impact on the <u>whole</u> system; e.g. if locks are established for a too long period.

Thus, NAV client PC which have to handle a **huge** transaction volume should consider this hardware configuration:

Component	Recommendation	Comment
CPU	1 x DualCore 1 x QuadCore	
RAM	4 GB	Regard OS limitations!
Network	Gigabit Ethernet	
Operating System	Windows XP or higher Vista cannot be used with older NAV versions!	Use Server OS if more RAM is necessary
Disk Subsystem	C:\	Operating System, Swap-File, NAV Client Program
	D:\	Misc, NAV *TempFilePath*

The defined "**Record Set**" size (see below "*Database Configuration*") could have major impact on the required hardware and performance!

The NAV/SQL Performance Field Guide
Version 2009

> Citrix/Terminal Server Hardware Environment

According to the "NAV Hardware Sizing Guide" a Citrix/TS server should be sized on basis of this formula:

- Calculate 10 to 15 users per CPU (logical, Cores) – rather calculate "conservative"
- Calculate 64MB of RAM per user (assuming 32MB "Object Cache")
- Calculate 1 or 2 GB RAM for the OS
- Calculate 500MB disks pace per user (SCSI or SAS, 15k rpm, RAID1)
- Gigabit Network

Example for 30 Users:

CPU: 30 / 10 = 3 CPU = 1 QuadCore
RAM: 30 x 64MB = 1920MB (2GB) + 2GB OS = 4GB
Disk: 30 x 500MB = 15000MB (15GB)

So you might end up with a server like this for e.g. 20 to 30 users:

Component	Recommendation	Comment
CPU	1 x QuadCore	
RAM	4 GB	Regard OS limitations!
Network	Gigabit Ethernet	
Operating System	Windows XP or higher Vista cannot be used with older NAV versions!	Use Server OS if more RAM is necessary
Disk Subsystem	C:\ (RAID1 [2 x HDD])	Operating System, Swap-File, NAV Client Program
	D:\ (RAID1 [2 x HDD])	Misc, NAV *TempFilePath*

Again, the defined "**Record Set**" size (see below "*Database Configuration*") could have major impact on the required hardware and performance!
With Citrix/TS it could be feasible to set the "Record Set" to a lower value than with "Fat-Clients".

Have in mind that this only applies to the NAV requirements! Other applications – like Office etc. – need system-resources, too. You should consider to add more hardware, or to reduce the number of users per Citrix/TS.

> Middle Tier Environment (NAV 2009)

Actually the 3-tier architecture of NAV 2009 is somewhat similar to the Citrix/TS architecture; hence the hardware requirements of the "*Middle Tier Server*" (aka "*Dynamics NAV Server*" or "*Application Tier*") are similar.

You should consider 1 dedicated (NAV only!) "*Middle Tier*" server per about 50 concurrent users (maximum – depends on transaction volume!):

Component	Recommendation
Architecture	32bit
CPU	1 x QuadCore
RAM	4 GB
Network	Gigabit Ethernet
Operating System	Windows Server 2003 or higher Standard
Disk Subsystem	C:\ (RAID1 [2 x HDD])
	D:\ (RAID1 [2 x HDD])

You should never run a "Middle Tier" service directly on the database/SQL server – except the DB server has <u>plenty</u> of hardware resources and you are running just few users.

> Used Version and Edition of the Operating System (OS)

Again, there are good advices in the "*Hardware Sizing Guide*". It is recommended to use an OS which has no limitations regarding the number of CPU used or the amount of RAM supported, to have a highly scalable system which could grow with the demands.

>> Memory – 32bit Architecture

In a 32bit system 2^{32} Bytes = **4.294.967.296 Bytes** (= 4 Giga Bytes) could be directly addressed by the Operating System. That's it. Addressing higher memory spaces requires additionally enabled features.

By default the OS splits these 4GB 50/50: 2 GB for the OS, 2 GB for applications. Hence, out-of-the-box a SQL Server cannot use more than 2GB RAM.

This 50/50 split could be overruled by setting the **/3GB** switch in `boot.ini`! If this is enabled, the OS will use 1GB and allows applications to allocate up to 3GB RAM.

To address higher memory spaces – beyond 4GB - some OS support a **PAE** (*Physical Address Extension*, an Intel chipset feature) feature which could be also enabled in `boot.ini` by adding the switch **/PAE**.

The boot.ini file is only available until Windows Server 2003. Since *Windows Server 2008* the boot-information is stored in the "**Boot Configuration Data**" (BCD) store (see http://technet.microsoft.com/en-us/library/cc721886(WS.10).aspx about details). The pendent to the /3GB switch could be set via the **bcdedit.exe** tool:

```
bcdedit /set IncreaseUserVA 3072
```

(3GB = 3 x 1024MB = 3072)

To undo/reset the changes:

```
bcdedit /deletevalue IncreaseUserVA
```

To make SQL Server use the higher memory space the **AWE** (*Address Windowing Extensions*) feature needs to be enabled (SQL Server site).

Before enabling AWE in SQL Server it has to be assured, that the "*Local Security Policies*" allow to "**Lock Pages in Memory**" for the account used by the SQL Server service (see below)!
AWE has to be enabled on SQL Server site; either by changing the properties of the instance (SQL Server 2005/2008) or via TSQL:

```
EXEC sp_configure 'show advanced options', '1'
RECONFIGURE
EXEC sp_configure 'awe enabled', '1'
```

When AWE is used **the Minimum** and **Maximum SQL Server Memory** has to be defined:

> *Minimum* = 0
> *Maximum* = *Physically available RAM – 1 [or 2] GB*

When addressing more than 16GB with SQL Server, the OS requires more memory, thus the /3GB switch has to be removed then.

All internal system data, execution plans, resource management etc. in SQL Server is managed/cached within the "*directly addressable memory space*", thus within the 2 to 3 GB of the lower memory space. All memory beyond the 4GB space are **data-cache** only!

Hence, 32bit systems have a natural limit in processing a certain transaction volume.

The NAV/SQL Performance Field Guide
Version 2009

Switch Configurations:

SQL Server 2000 (32 bit)

SQL	OS	RAM	/3GB	/PAE	AWE
STD	Windows 2003 STD	<= 2GB	✗	✗	✗
EE	Windows 2003 EE	> 2GB <= 4GB	✓	✓	✗
EE	Windows 2003 EE	> 4GB <= 16GB	✓	✓	✓
EE	Windows 2003 EE	> 16GB	✗	✓	✓

SQL Server 2005/2008 (32 bit)

SQL	OS	RAM	/3GB	/PAE	AWE
STD/EE	Windows 2003 STD	<= 4GB	✓	✓	✗
STD/EE	Windows 2003 EE	> 4GB <= 16GB	✓	✓	✓
STD/EE	Windows 2003 EE	> 16GB	✗	✓	✓

STD = Standard Edition
EE = Enterprise Edition
With *Windows Server 2008* use the **BCD** feature instead of /3GB

>> Memory – 64bit Architecture

With 64bit systems 2^{64} Bytes = **18.446.744.073.709.551.616** Byte (= 18 Exa Bytes) can be <u>directly</u> addressed. Thus, no additional features are required to use high memory spaces, so no /3GB /PAE or AWE is required.

With 64bit systems all available RAM could be used as data- and procedure cache, and to manage various kinds of internal SQL data.

As sufficient memory is crucial for the SQL Server's performance, 64bit machines should be obligatory nowadays.

Highly recommended!

<u>Caution</u>: With **64bit** systems it might be also necessary – depending on the used SQL Server version/edition: you need either **SQL 2005 SP3 Update 4** or **SQL 2008 SP1 Update 2** or higher – to grant the "**Lock Pages in Memory**" right to the Windows account used by the SQL Server service; else performance could be degraded.

Important! Read this first: http://support.microsoft.com/kb/918483
                                http://support.microsoft.com/kb/970070

To assign the "*Lock Pages in Memory*" user right, follow these steps:

1. Click **Start**, click **Run**, type **gpedit.msc**, and then click OK; the **Group Policy** dialog box appears.
2. Expand **Computer Configuration**, and then expand **Windows Settings**.
3. Expand **Security Settings**, and then expand **Local Policies**.
4. Click **User Rights Assignment**, and then double-click *Lock Pages in Memory*.
5. In the **Local Security Policy Setting** dialog box, click **Add User or Group**.
6. In the **Select Users or Groups** dialog box, add the account that has permission to run the SQL Server service, and then click OK.
7. Close the **Group Policy** dialog box.
8. **Restart** the SQL Server service.

(taken from the KB article)

Then enable the **Trace Flag 845** (add Startup-Parameter -**T845**).
Restart the SQL Server service.

>> Processors

It is recommended to use currently state-of-the art CPU; no cheap light-versions or other stuff.
Have in mind that **Dual Core** is not the same as **Dual CPU**, even though both have its advantages (last but not least when licensing SQL Server by CPU):

- ✓ A **Dual Core** is actually one physical processor split into 2 logical processors or "cores". This processor has one *Processor Queue* (could be a disadvantage). The advantage is, that both cores are dynamically sharing one *L2 Cache*.

- ✓ A **Dual CPU** means to have two physical processors, each with a single core. Each CPU has its own *Processor Queue* and *L2 Cache*. The disadvantage here could be, that if the *L2 Cache* is full, the CPU slows down.

It could be very feasible to use a combination of both, e.g. to install two Dual Core CPU, giving 2 physical and 4 logical processors!

It is recommended to use 1 physical CPU per 100 processes (or users); for example with 150 users you should have 2 physical CPU, e.g. 2 QuadCores (= 8 logical CPU).

Remark:
When buying SQL Server with "*Per Processor License*" only physical CPU are charged!

Hyperthreading

According to "logical processors", some OS and/or BIOS support the feature "*Hyper-Threading*" (HT). HT is a frequently misunderstood feature:

HT does not increase or expand the CPU **power** – HT is used to put more **load** on the CPU!

This is how *Hyper-Threading* works (simplified):

HT actually "fakes" the number of available CPU; for example, if 2 CPU are installed, the OS will show 4. It is obvious that there cannot possibly be some kind of miraculous multiplication of CPU power – it is still just 2 physical CPU.
As the OS "thinks" there are more CPU it will create more Threads than before (without HT). Means, when previously a process was handled with 2 Threads – one per CPU – the now the same process generates 4 Threads.

STRYK System Improvement
Performance Optimization & Troubleshooting

Hence, each physical CPU has to process 2 Threads; the load on the physical CPU is doubled.

Normal Mode – non hyper-threaded:

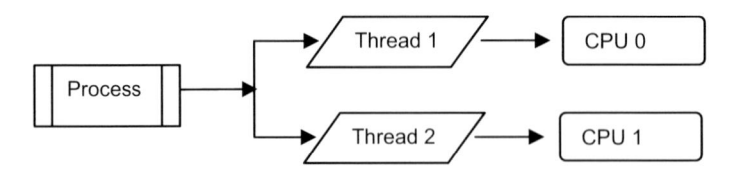

```
Process → Thread 1 → CPU 0
        → Thread 2 → CPU 1
```

Hyper-threaded:

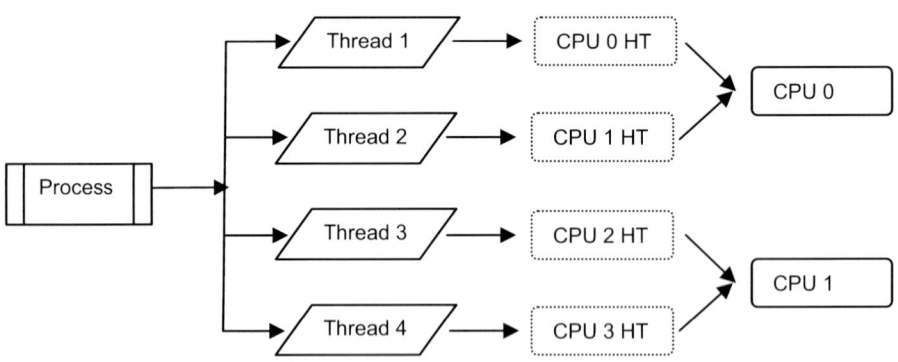

```
Process → Thread 1 → CPU 0 HT ┐
                              ├→ CPU 0
        → Thread 2 → CPU 1 HT ┘
        → Thread 3 → CPU 2 HT ┐
                              ├→ CPU 1
        → Thread 4 → CPU 3 HT ┘
```

Generally the benefit of this is to assign more CPU time to a process, which should increase the performance up to 30%.

Unfortunately, HT does not work properly with SQL Server (browse the wwweb for technical details).

Thus, *Hyper-Threading* is not recommended with SQL Server – **real physical CPU power** gives better performance!

> Used Version and Edition of the SQL Server

When selecting the right version ([2000 or] 2005 or 2008) and edition (Standard, Enterprise, etc.) besides the desired **features** the capacity of the SQL Server is important. As mentioned above, the required capacities depend on the **expected data/transaction volume** to be processed.

The following gives an overview about CPU and RAM capacities; further details can be seen on the related web-pages at http://www.microsoft.com/sql/

Microsoft SQL Server 2000 – Maximum number of supported CPU and RAM

Feature	Enterprise Edition	Standard Edition	Personal Edition	Desktop Engine
CPU	32	4	2	2
RAM	64GB	2GB	2GB	2GB
64-bit Support	Yes	No	No	No
Database Size	unlimited	unlimited	unlimited	4 GB

Microsoft SQL Server 2005 – Maximum number of supported CPU and RAM

Feature	Enterprise Edition	Standard Edition	Workgroup Edition	Express Edition
CPU	unlimited	4	2	1
RAM	unlimited	unlimited	3 GB	1 GB
64-bit Support	Yes	Yes	WoW	WoW
Database Size	unlimited	unlimited	unlimited	4 GB

Microsoft SQL Server 2008 – Maximum number of supported CPU and RAM

Feature	Enterprise Edition	Standard Edition	Workgroup Edition	Express Edition
CPU	unlimited	4	2	1
RAM	unlimited	unlimited	4 GB	1 GB
64-bit Support	Yes	Yes	Yes	Yes
Database Size	unlimited	unlimited	unlimited	4 GB

STRYK System Improvement
Performance Optimization & Troubleshooting

Think careful when picking a SQL Server edition!

Each *edition* is a separate product/SKU.
Thus, when purchasing – for example – a *SQL Server 2005* **Standard** edition it is possible to **upgrade** to *SQL Server 2008* **Standard** edition (depending on any contract with Microsoft as "Select" or "Software Assurance" etc.).

But it is not possible to "upgrade" from any **Standard** edition to **Enterprise** edition; this would mean to (re)**buy** another SQL Server.

Calculate and Compare different pricing models:
- Server & CAL
- Per Processor (physical CPU, not logical CPU)
- NAV Runtime (highly recommended with dedicated NAV systems and few users!)

Regard your existing licensing/support contracts with MS and ask your MS Partner about current promotions etc..

Some "**Enterprise**" features which could be important regarding performance and should be discussed when selecting the SQL Server edition:
- Online Indexing (2005/2008)
- Asynchronous Mirroring (2005/2008)
- Indexed Views (2005/2008)
- Data Compression (2008)
- Backup Compression (2008)
- Resource Governor (2008)

See **Appendix B** for existing versions of MS Dynamics NAV and SQL Server.

The NAV/SQL Performance Field Guide
Version 2009

> Configuration of SQL Server and Database

SQL Server provides "gazillions" of features to be configured properly. The following describes the most important settings where changes affect performance.
Whatever is not mentioned in the following list I recommend using the **default values**. This means you should be aware of the **default values** of the settings!

Further it is assumed the server is a dedicated SQL Database Server running only one instance of SQL Server which is exclusively used by NAV!

Never abuse a database server as file- or print/spool-server, or domain-controller, etc.!
Never run other heavy services or applications on the database server!

>> SQL Server Instance Settings

The following refers to **SQL Server 2005/2008**. With *SQL 2000* the settings are similar, major differences are mentioned in the "Comment" column.

SQL Server Configuration Manager

Property	Default	Recommended	Comment
Startup parameter	n/a	-T4616	Required since NAV 4.0
			Not required anymore since NAV 2009 SP1
Startup parameter	n/a	-T4119	Optional; only to prevent specific problems (see below)
Startup parameter	n/a	-T1204 -T1222	Optional; to write Deadlock info into the SQL Error Log
			SQL Server 2000:
			-T1204
			[-T1205 optionally; do not use if too much data is written in the Error Log then]
			-T3605

The "*Startup parameters*" above have to be added to the existing ones! Refer to the "books Online" about further information.

STRYK System Improvement
Performance Optimization & Troubleshooting

SQL Server Management Studio

Property	Default	Recommended	Comment
Memory			
AWE Enabled	No	Yes	Only on 32bit servers with more than 4GB RAM (see above)
Min. Server Memory	0	0	On a dedicated server it is not necessary to increase the minimum. If running multiple instances or other services on the server, then this value might be increased. Caution: if this value is too high the OS might encounter problems with starting other services!
Max. Server Memory	2147483647	RAM – 1 [or 2] GB	Must be set when using AWE
Processors			
Processor Affinity	Yes	Yes	On a dedicated server all available CPU should be used; the affinity masks set as default. If the OS is processing other tasks/services this is usually done with CPU 0. If there is high pressure (caused by OS) on CPU 0 then it could be feasible to exclude it from the SQL Server Instance. Network I/O is partly managed on the <u>last</u> CPU; if there is too high pressure, too, then it also might be excluded.
I/O Affinity	Yes	Yes	
Max. Worker Threads	0 SQL 2000: 255	0 SQL 2000: 255	With default 0 the SQL Server creates worker threads automatically on basis of the available CPU. | CPU	32bit	64bit
---	---	---
<= 4	256	512
8	288	576
16	352	704
32	480	960	 If the number of connection **exceeds** the "*max. worker thread*" value, then SQL Server has to perform "**thread pooling**" which might degrade performance. In this case the value has to be increased manually. Recommended Maximum: 32bit: 1024 64bit: 2048

SQL Server Management Studio

Property	Default	Recommended	Comment
Processors			
Priority Boost	No	No	Not necessary on dedicated servers. Caution: This boost option sets the process priority from 7 (normal) to 13 (high). This is actually higher than network processes (prio 10), which could mean, that on a heavy load on the server there will be "drop-outs" in the network – performance decreasing.
Lightweightpooling (NT Fiber Mode)	No	No	Enabling it could solve "*Context Switch*" problems, but actually this feature does not work with NAV and Windows Authentication.
Database Properties			
Standard Fill-Factor	0	0	Caution: NEVER apply a fill-factor here!
Advanced			
Max. Degree of Parallelism	0	1 Or 50% CPU	The default value (0) allows the SQL Server to involve all available CPU to create parallel "Execution Plans"(QEP) if necessary. Basically NAV queries are too simple to be "parallelized". But with **SIFT systems** (<= NAV 4.0) it could happen that processes are indeed split into multiple threads. Unfortunately this often leads to blocking issues – the multiple threads are locking each other – which could end up in "Deadlock" situations (= "suicide process"). By setting this value to 1 no parallel QEP are generated, thus no self-blocking could happen. With **VSIFT systems** (>= NAV 5.0) this risk is minor or void, parallelism could be allowed. But even in this case it is recommended not to include all CPU, but maybe just half of them. In this case no process could occupy all CPU at one time; the risk of one process *overloading* the system is smaller.

STRYK System Improvement
Performance Optimization & Troubleshooting

>> Database Settings

SQL Server Management Studio

Property	Default	Recommended	Comment
Files			

Auto. Growth	10%	250MB	With a relative growth rate the "chunks" for each expansion get bigger and bigger (= *progressive growth*), thus a **fixed value** is preferable (= *linear growth*).
			Actually, "Auto Growth" should just be used as some kind of "emergency" expansion, to avoid running into a db-halt. Ideally, an administrator takes care about the correct file sizes.
			
			As a "rule of thumb" one could say, the database should have **free space** of **20% to 25%**, or **1,5 times the size of the largest table** (data and indexes). This space would be used e.g. when re-indexing to avoid the creation of too many "*Overflow Pages*".
			
			Especially the "Transaction Log" should net "auto grow" frequently, as with every expansion more "**Virtual Transaction Logs**" (VLogs) are created which cause more administrative overhead.
			
			Check the number of "**Virtual Logs**" (VLogs) by executing this TSQL:
			
			`use [Navision]-- change db name`
			`go`
			`dbcc loginfo`
			
			The number of lines shown represents the number of VLogs.
			To reduce the number of VLogs one could proceed like this:
			1. Truncate the TLog (e.g. backup)
			2. Shrink the TLog to a <u>minimal</u> size
			3. Adjust the "*Initial Size*" of the TLog to the required <u>maximum</u> size

SQL Server Management Studio

Property	Default	Recommended	Comment
Options			
Auto. Close	False	False	Never let the server decide when to close the db
Auto. Shrink	False	False	Never use this – the NAV db will grow steadily, "Auto. Shrink" might cause additional fragmentation of indexes which would degrade performance.
Auto. Update Statistics	True	False	The statistics which are automatically created are insufficient and cause additional load on the system (delaying read and write transactions); in some cases these stats could be responsible for some errors in the NAV development. Sufficient stats should be created/maintained by **SQL Agent Jobs** or **Maintenance Plans**.
Auto. Update Statistics Asynchronously	False	False	
Auto. Create Statistics	True	False	
Page Verify	Checksum	Checksum	With some older SQL 2000 db this value might be "**Torn Page Detection**" (TPD). This should be changed, as TPD is degrading the write performance. SQL 2000: "**Torn Page Detection**" should be disabled and be compensated by **SQL Agent Jobs** or **Maintenance Plans** (DB Integrity Checks)

Once the "Auto Stats" features have been disabled it is necessary to remove the previously generated auto-statistics:

```
DECLARE @id INT, @name VARCHAR(128), @statement NVARCHAR(1000)
DECLARE stat_cur CURSOR FAST_FORWARD FOR
   SELECT [id], [name] FROM sysindexes
   WHERE (indexproperty([id], [name], N'IsStatistics') = 1)
     AND (indexproperty([id], [name], N'IsAutoStatistics') = 1)
     AND (isnull(objectproperty([id], N'IsUserTable'),0) = 1)
   ORDER BY object_name([id])
OPEN stat_cur
FETCH NEXT FROM stat_cur INTO @id, @name
   WHILE @@fetch_status = 0 BEGIN
      SET @statement = 'DROP STATISTICS [' +
                       object_name(@id) + '].[' + @name + ']'
      BEGIN TRANSACTION
         PRINT @statement
         EXEC sp_executesql @statement
      COMMIT TRANSACTION
      FETCH NEXT FROM stat_cur INTO @id, @name
   END
CLOSE stat_cur
DEALLOCATE stat_cur
```

After this sufficient stats have to be generated using the "**sp_updatestats**" and "**sp_createstats**" procedures (see below about "*Index Statistics*").

>> NAV Client Settings 

Some of the database settings can also be done from NAV client site.

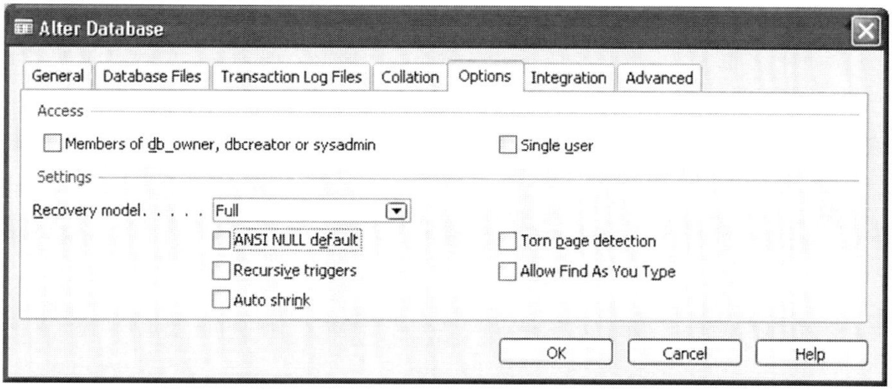

In addition to the issues mentioned above, it is recommended to disable the "**Find As You Type**" feature. If this is used, each keystroke sends a query to the server! These queries could use "wild-cards" – e.g. when using "Part of Field" - , causing mostly bad execution plans as "*Clustered Index Scans*", which is performing very poor.

Dynamics NAV 2009 (not shown in screenshot):
The property "**Enable for MS Dynamics NAV Server**" should only be used when running the **3-Tier architecture**. When running NAV 2009 Classic Client only, this should be disabled as it might degrade performance.

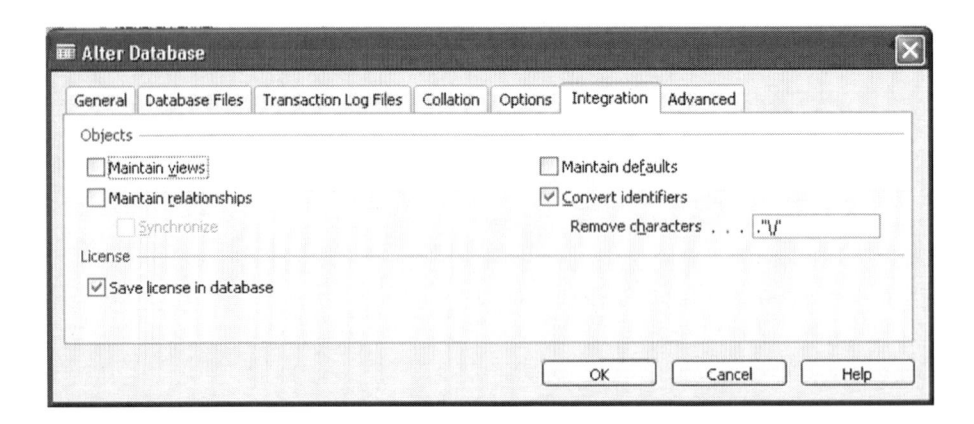

STRYK System Improvement
Performance Optimization & Troubleshooting

"**Maintain Views**" would create multiple SQL Server Views (per table, per language layer, throughout the whole system), which are using the *Field Captions* instead of *Field Names*, so e.g. other applications can query data from these Views using local-language names. Should be disabled.

"**Maintain Relationship**" would establish "*Foreign Key*" (FK) relations between tables on SQL Server. This has advantages regarding data-consistency, but is severely decreasing performance. Should be disabled.

"**Maintain Defaults**" would create – and maintain - default constraints for each field in the system, which could be used by external applications. Should only be enabled if required by external applications.

The basic problem with these settings is, that they have impact on the whole system – it's ON or OFF. It is more feasible, to implement the desired features – Views, Foreign Keys, Default – when and where they are **really required**.

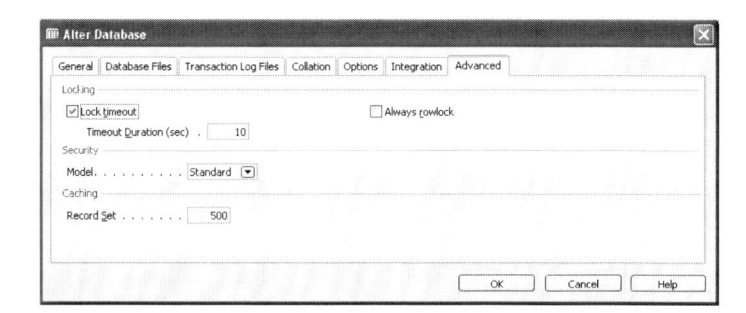

If a "**Lock Timeout**" is defined, a transaction would be cancelled if it is blocked longer than the **Duration** specifies.
This could be problematic for long batch transactions or unattended processes, so have in mind that the "Lock Timeout" could be enabled/disabled with the C/AL command `LOCKTIMEOUT(Boolean)`.
For example, disabling the "Lock Timeout" when running a process on the NAS:

`LOCKTIMEOUT(GUIALLOWED);`

If "**Always Rowlock**" is enabled, C/SIDE will add the optimizer-hint "ROWLOCK" to the queries sent to the SQL Server. Locking mechanisms in SQL Server are quite complex and cannot be fully discussed here (see below "**Locking Mechanisms**").

Normally, SQL Server will always start locking on row-level and then escalate if necessary. **The ROWLOCK hint prevents/delays this escalation**. In most cases, the administration of many row-locks is performing worse than having an escalated range-lock, thus performance is decreasing.

As locks are primarily maintained within the *directly addressed memory space*, ROWLOCK could cause more problems with **32bit systems** (4GB directly) than with 64bit (18 Mio TB directly) – mostly 64bit systems with a sufficient amount of RAM have no problems with "*Always Rowlock*".

Hence, "*Always Rowlock*" could increase the parallelism of the system, but also could cause problems.

STRYK System Improvement
Performance Optimization & Troubleshooting

The "**Record Set**" size defines the max. number of records queried per call (using the FINDSET method). E.g. with a setting of 500 (standard), a result-set which includes 500 records is queried with the first call, then further queries will select packages (each with less records) until all records are retrieved.

Actually, the idea behind `FINDSET` is this:
The "old" `FIND('-')` created a cursor containing all records of the filter – regardless if it was 1 record or 1 million. As all data is sent to the NAV Client, huge cursors could overload the client's RAM capacities, degrading performance.
`FINDSET` was introduced to create smaller cursors, by creating multiple smaller "*packages*" of the result-set, not overloading the client's memory. A potential problem is, that the more calls ("packages") are performed, the more the performance could be decreased.

Normally a "**Record Set**" size of 500 is quite OK for most NAV "Fat Clients".
On Citrix/TS environments it could be feasible to reduce this value, e.g. to 200 or 300. With NAV 2009 the default value was even set to 50. The reason is that here one machine – Citrix/TS Server or Middle Tier - has to process the cursors from multiple user-processes!
Hence, when running NAV 2009 Classic Client only – no Middle Tier – the "Record Set" size could be increased, again.

It is recommended to set the "**Security Model**" to "*Standard*", except there is the specific need for the "*Enhanced*" model (MS provides a document comparing both models).
With "*Standard*" the **User-Synchronization** process is much faster, as actually just the Application Role `$ndo$shadow` is used.

With "*Enhanced*" NAV generates an Application Role (`$ndo$ar$...`) for each defined Login. When "*Synchronizing*" all NAV access-rights are transferred into each role, which could be very time consuming!

To change the "*Security Model*" the database has to be in "**Single User**" mode.

Caution: With "*Standard*" security the database-users which are ONLY defined on SQL Server site (e.g. to access the DB with other applications etc.) and NOT within NAV as Database- or Windows-Login are deleted when running the "*Synchronization*"!

>> Example TSQL for Configuration
(SQL Server 2005/2008)

```
/* Adjust settings */

-- AWE Configuration
EXEC sp_configure 'awe enabled', '1'
GO

EXEC sp_configure 'max server memory (MB)', '8000'
GO

-- Parallelism
declare @dbversion int, @cpu int, @maxdop int
if exists(select top 1 null from sys.tables
          where [name] = '$ndo$dbproperty')
  select @dbversion = [databaseversionno]
  from [dbo].[$ndo$dbproperty]
else
  set @dbversion = 999
select @cpu = [cpu_count] from sys.dm_os_sys_info
if @dbversion < 95    -- SIFT
  set @maxdop = 1
else begin            -- VSIFT
  set @maxdop = @cpu / 2
end

-- Allow shell-command execution
EXEC sp_configure 'xp_cmdshell', '1'
GO

-- End Configuration
RECONFIGURE WITH OVERRIDE
GO
```

```sql
/* NAV Database Configuration */
-- General Settings
ALTER DATABASE [Navision] -- change database name here
SET AUTO_CLOSE OFF
   ,AUTO_CREATE_STATISTICS OFF
   ,AUTO_SHRINK OFF
   ,AUTO_UPDATE_STATISTICS OFF
   ,AUTO_UPDATE_STATISTICS_ASYNC OFF
   ,RECOVERY FULL
   ,TORN_PAGE_DETECTION OFF
   ,PAGE_VERIFY CHECKSUM
GO


USE [Navision] -- change database name here
GO

DECLARE @file VARCHAR(250), @statement NVARCHAR(1000)
DECLARE dbfile_cur CURSOR FOR SELECT [name] FROM sysfiles
OPEN dbfile_cur
FETCH NEXT FROM dbfile_cur INTO @file
WHILE @@fetch_status = 0 BEGIN
  SET @statement = N'ALTER DATABASE [' + db_name() + '] MODIFY FILE (NAME = ''' + @file + ''', FILEGROWTH = 250 MB)'
   EXEC sp_executesql @statement
   FETCH NEXT FROM dbfile_cur INTO @file
END
CLOSE dbfile_cur
DEALLOCATE dbfile_cur
GO

-- NAV specific settings
UPDATE dbo."$ndo$dbproperty"
SET quickfind = 0
   ,locktimeout = 0
   ,hardrowlock = 0
GO
```

> Optimizing "tempdb"

Usually the "*tempdb*" is just used moderately with NAV. But when dealing with large result-sets, containing lots of records, there could be remarkable "pressure" on this db, e.g. if heavy sort operations are required.

In this case the first thing should be to move the "*tempdb*" to a **dedicated hard-drive** (as shown above).

Then also the "**Auto. Growth**" properties of the database files should be set to fixed values and the "**Auto. Statistics**" features should be disabled (same as with NAV database).
Make sure the **initial file size** is as big as the expected maximum size to avoid "*Auto Growth*".
Use the "*Simple*" **Recovery Model**.

It is feasible to split the "**tempdb**" into multiple data-files; a rule of thumb is "*one file per CPU*" but this should be adjusted depending on the local requirements.
Too many files on the same physical drive might exceed the capacities of a disk-controller, depending on its technology!

I recommend to split the "tempdb" like this:

CPU	Number of files
2	1 to 2
4	2 to 4
>= 8	4 to 6

See also http://msdn.microsoft.com/en-us/library/ms175527.aspx

STRYK System Improvement
Performance Optimization & Troubleshooting

Fixing the Trouble – Erasing problems

Once the fundamental setup and environment is sufficient it could be started to optimize different issues in the system. There are some basic "mistakes" or "malfunctions" in NAV which have to be fixed. Then the "fine tuning" of the system could be performed.

These essential issues to be improved are:

- Structure of Indexes in NAV
- Structure of SIFT Indexes in NAV
- Structure of VSIFT Views in NAV
- Design of C/AL Code

> Structure of Indexes in NAV

Optimizing the Index structure – besides the C/AL design - in NAV will give the most "boost" in performance. A prerequisite here is a basic understanding about keys and indexes.

Key = a **constraint** for a record or a field

The most important key is the **Primary Key** (PK), which is a constraint that defines the *uniqueness* of a record. A table could have only one PK. A **Foreign Key** (FK) is a constraint of a field which is related to the PK of another table.

Index = an internal, **balanced tree structure** (B-Tree) to retrieve data

In SQL Server two types of indexes are distinguished: The **Clustered Index** (CI) is an index which defines the physical order of the records. A table can have only one CI. The "leaf nodes" of a CI actually include the data, they represent the table. All other indexes are called **Non-Clustered Indexes** (NCI).

Balanced Tree (B-Tree) scheme:

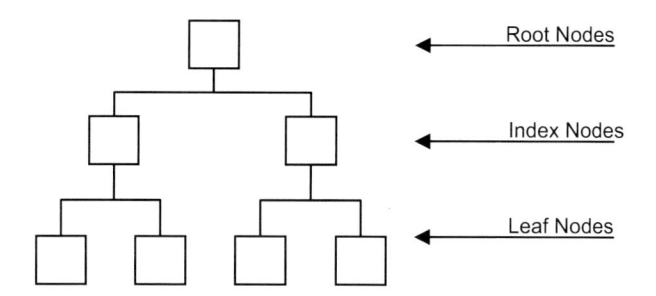

Root Nodes

Index Nodes

Leaf Nodes

Key and Index handling is quite <u>different</u> in SQL Server than with the "native" NAV database server (FDB). What are called "Secondary Keys" (SK) in FDB are actually indexes. Keys and indexes in FDB are used for *sorting, SIFT handling and data-retrieval*. SQL Server uses indexes exclusively for *data-retrieval*.

STRYK System Improvement
Performance Optimization & Troubleshooting

Non-Clustered Index:

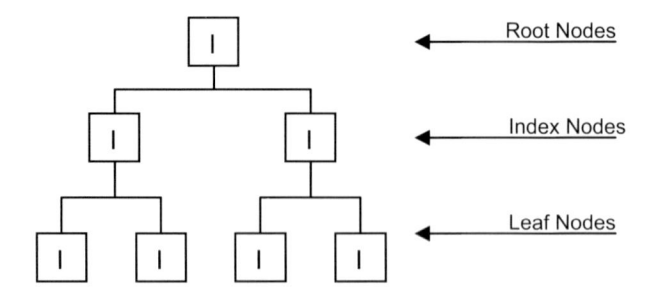

```
              [ I ]  ← Root Nodes
             /     \
          [ I ]   [ I ]  ← Index Nodes
          /  \    /  \
        [I] [I] [I] [I]  ← Leaf Nodes
```

All Nodes contain just the **indexed** value.

Clustered Index:

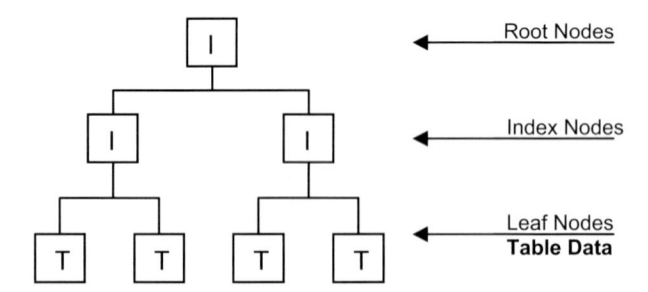

```
              [ I ]  ← Root Nodes
             /     \
          [ I ]   [ I ]  ← Index Nodes
          /  \    /  \
        [T] [T] [T] [T]  ← Leaf Nodes
                         Table Data
```

Root- and Tree- Nodes contain the **indexed** values. The Leaf Nodes additionally contain the **table data / records**.

Since SQL 2005 so called "**Included Columns**" are supported with indexes (NCI). Here columns/fields from the table could be added only to the *Leaf Nodes* of an indexes; the "Included Columns" are not stored in the B-Tree.

Non-Clustered Index with Included Columns:

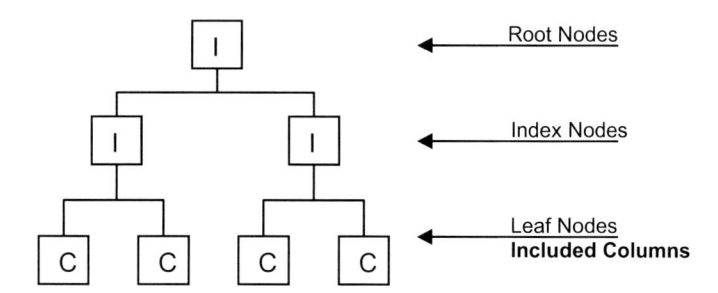

```
                    [ I ]           ←  Root Nodes
                   /     \
                [ I ]   [ I ]       ←  Index Nodes
                /  \     /  \
              [C] [C] [C] [C]       ←  Leaf Nodes
                                       Included Columns
```

Root- and Tree- Nodes contain the **indexed** values. The Leaf Nodes additionally contain the **included columns**.

Hence, a Non-Clustered Index could include an *excerpt* of the table data. As these includes are only attached to the leaf node level, the index tree is still small, but additionally required information could be added, e.g. to avoid looking up the Clustered Index.

Caution: Adding all table fields as Included Column would actually mean to virtually copy the table (in a different sorting order)! Hence, the index size would be remarkably increased!

STRYK System Improvement
Performance Optimization & Troubleshooting

When indexes are defined in NAV, C/SIDE "translates" them into the SQL Server specific pendants:

C/SIDE	SQL Server
Primary Key	Primary Key Clustered Index
Secondary Key	Non-Clustered Index (UNIQUE)
Table Relation (Property)	Foreign Key* *) if db property "*MaintainRelations*" is enabled

When C/SIDE is creating a NCI it sets the flag UNIQUE. If this flag is set, SQL Server checks if the index value exists just once; it's a similar constraint as the PK constraint. To grant this uniqueness, C/SIDE adds all PK fields (if not already part of the index) to the NCI.

Example:

Table 18 Customer
The PK "*No.*" generates a PK/CI "*No.*"
The SK "*Search Name*" is created as NCI "*Search Name*", "*No.*" UNIQUE

This actually shows one of the fundamental **insufficiencies** of the NAV index creation:

The NCI is created **UNIQUE**. So just an index "*Search Name*" would cause severe logical problems as it would mean, that for example just one customer named "Smith" could be inserted. So the PK field has to be added to grant this uniqueness. The result is, that SQL Server always has to check if the index value is UNIQUE, but there is no chance at all to be not unique (due to the PK which is part of the index)!

Hence, the indexes created by C/SIDE cause a lot of administrative **overhead** for SQL Server which is not necessary, and the **size** of the index is unnecessarily increased, which slows down read- and write-transactions!

So, when optimizing indexes these measures have to be taken:

- Reduction of the **number** of Indexes on SQL Server site
- Optimizing the **order** of fields, especially of the Clustered Index
- Remove **UNIQUE** flag and added **PK fields** from the Non-Clustered Indexes
- Define optimized **Fill-Factors** for the indexes
- [Move indexes to a dedicated **File-Group**]

>> Reduction of the number of Indexes

In C/SIDE there are often very similar indexes defined within a table; mostly, because the index is just used for a different sorting. As mentioned, SQL Server just needs the indexes for data-retrieval, so all unnecessary indexes should be removed from SQL Server site by setting the "**MaintainSQLIndex**" to FALSE.
Maintained indexes should not be too granular (means too many fields) as SQL Server prefers small but efficient indexes.

The "*NAV SQL Server Option Resource Kit*" (or "*Database Resource Kit*") provides some tools to get a convenient overview about the index structure of tables. Tables with a high "**Costs per Record**" (relation of data insertion to index & SIFT insertion) and a high number of Records should be optimized first.

The indexes of a table could be analyzed with TSQL

```
EXEC sp_helpindex [ @objname = ] 'name'
```

With SQL Server 2005/2008 the actual index usage could be displayed:

```
SELECT * FROM sys.dm_db_index_usage_stats
```

Caution: Have in mind that this statistic (as other DMV) is reset when restarting the SQL Server service! To safely deactivate indexes it is crucial to have a representative statistic!

STRYK System Improvement
Performance Optimization & Troubleshooting

>> Optimizing the order of fields

SQL Server has actually opposite requirements for indexes than FDB. While in FBD e.g. an option field in the beginning like "Document Type" is quite feasible, for SQL Server this is very sub-optimal. SQL Server indexes should begin with highly **selective** fields, e.g. like "*Document No.*".

Explanation of "**Selectivity**" with a simplified example for table "*Sales Line*":

PK/CI: *"Document Type", "Document No.", "Line No."*

Assuming there were 4 different "Document Types" used: *Quote, Order, Invoice, Credit Memo*

There are 100 records, 25 per "*Document Type*", "*Document No.*" from 1 to 25 per "*Document Type*"

So the selectivity of "*Document Type*" is **25%**; e.g. when filtering on "Order" the result-set is 25 records of the whole volume.

The field "*Document No.*" has a selectivity of **4%** as each number just occurs once per "*Document Type*"; e.g. when filtering on 1 the result-set is just 4 records.

The Selectivity of index fields could be investigated with TSQL

```
DBCC SHOW_STATISTICS ( table , target )
```

Before	After
DBCC SHOW_STATISTICS ('CRONUS$Sales Line',	'CRONUS$Sales Line$0')

All density	Average Length	Columns	All density	Average Length	Columns
0,25	4	Document Type	0,0001501051	6,950215	Document No_
0,0001685772	10,97277	Document Type, Document No_	1,51247E-05	10,95021	Document No_, Line No_
9,387908E-05	14,97277	Document Type, Document No_, Line No_	1,315322E-05	14,95021	Document No_, Line No_, Document Type

Especially for the CI it is <u>crucial</u> to have **optimal selectivity**!

Since NAV 4.00 SP1 it is possible to define different SQL Server site indexes than FDB indexes using the key properties "**SQL Index**" and "**Clustered**"!

IMPORTANT! Caution – be aware of this:

When NAV is sending a SELECT query to the server it includes an **ORDER BY** clause which is related to the current "**Key**" context (e.g. defined by a `SETCURRENTKEY` command in C/AL), if no certain "Key" is specified the Primary Key will be used for sorting.
If SQL Server picks an index to process the query optimally it normally regards the Filter – the **WHERE** clause – in first priority. If the sorting of the chosen index is very different to the defined sorting (ORDER BY) some problems could arise:
Either SQL Server is picking the Filter-Optimized Index and then it has to re-sort the records in the "*tempdb*"; or SQL Server is picking an suboptimal Index which matches better to the ORDER BY, which usually causes Index Scans.
Both problems are time consuming and degrading performance.

Hence, if the "*SQL Index*" is too different to the "*Key*" problems may occur.
"*SQL Theory*" and "*NAV Reality*" about indexes are not necessarily the same!

MS introduced several supposedly "optimized" *SQL Indexes* in NAV 5.0, which were completely removed in 5.0 SP1, as those indexes caused severe problems!

Several NAV versions are using so called "**Dynamic Cursors**" (DC) instead of "**Fast Forward Cursors**" (FFC).
DC are optimizing the "**Query Execution Plan**" (QEP) for the ORDER BY, thus "insufficient" SQL Indexes could create or worsen some problems!
FFC are optimizing for the WHERE clause, the "Key"/ORDER BY is less important.

For example:

```
SELECT * FROM "CRONUS$G_L Entry"
WHERE "Posting Date" = '2009-07-14 00:00'
ORDER BY "Entry No_"
```

With a FFC the index "**$1**" (*"G/L Account No.", "Posting Date"*) would be probably picked, giving good performance.
A DC would pick the index "**$0**" (*"Entry No."*), causing a problematic *index scan*.

Hence, the "**SQL Index**" property gives us a chance to define better than standard indexes but is has to be used very careful, else problems will be raised or worsened.

According to the NAV/SQL "*Cursor Handling*" it should be mentioned that with NAV 5.0 SP1 a "**Cursor Preparation**" feature has been introduced:

Here the procedure **sp_cursorprepare** is used to only generate an **Execution Plan** (QEP) for a query but <u>without</u> actually creating the cursor (which is done via **sp_cursoropen**). This allows the SQL Server to prepare an efficient QEP for the following "real" query – here the cached QEP could be used, saving CPU time.
The QEP generated by **sp_cursorprepare** are erased by **sp_cursorunprepare**.
As mentioned, the idea of this is to "help" the SQL Server compiling an optimized QEP.

Unfortunately this sometimes fails, in few cases the "Cursor Preparation" could generate "bad" QEP, thus creating/forcing a problem. If this happens, the preparation could be suppressed by adding a specific value to the "**diagnostics**" field in table "**ndodbproperty**":

```
USE [Navision]
GO
UPDATE [dbo].[$ndo$dbproperty]
SET [diagnostics] = [diagnostics] + 1048576
GO
```

To undo this change:

```
USE [Navision]
GO
UPDATE [dbo].[$ndo$dbproperty]
SET [diagnostics] = [diagnostics] - 1048576
WHERE [diagnostics] >= 1048576
GO
```

The default value of "diagnostics" is 0 (zero):

```
USE [Navision]
GO
UPDATE [dbo].[$ndo$dbproperty]
SET [diagnostics] = 0
GO
```

The NAV/SQL Performance Field Guide
Version 2009

Especially the order of the *Clustered Index* could have remarkable impact on **locking** issues, too!

Example:

Optimizing table 357 "*Document Dimension*"
PK: *Table ID, Document Type, Document No., Line No., Dimension Code*

This PK/CI starts with a somewhat non-selective field "*Table ID*". When NAV performs transactions with this table, e.g. during postings, it mostly queries the records of one "*Document No.*", both from Header and Line (e.g. *Sales Header* (36) and *Sales Line* (37). While posting, a range lock is established.

This is what happens when querying `"Table ID"` = 36|37, `"Document Type"` = 0, `"Document No."` = 1:

Original physical order

Table ID	Document Type	Document No.	Line No.	Dimension Code	Lock
36	0	1		Dim1	
36	0	1		Dim1	
36	0	2		Dim1	Block!
36	0	2		Dim1	Block!
36	1	1		Dim1	
36	1	2		Dim1	
36	2	1		Dim1	
37	0	1	10000	Dim1	
37	0	1	20000	Dim1	
37	0	2	10000	Dim1	
...	...	...	...	...	

A second process (user) queries `"Table ID"` = 36|37, `"Document Type"` = 0, `"Document No."` = **2**

In this scenario it is not possible to work with *Sales Quote (Document Type 0) No.2* as the records are blocked.

Page 73

STRYK System Improvement
Performance Optimization & Troubleshooting

Improved physical order with changed CI

Document No.	Document Type	Table ID	Line No.	Dimension Code	Lock
1	0	36		Dim1	▓
1	0	36		Dim1	▓
1	0	37	10000	Dim1	▓
1	0	37	20000	Dim1	▓
1	1	36		Dim1	
1	2	36		Dim1	
2	0	36		Dim1	
2	0	36		Dim1	
2	0	37	10000	Dim1	
2	1	36		Dim1	
...	...	...	...	...	

Here any locking problems are eliminated, other Documents can be processed.

With NAV 4.00 SP1 and higher you simply have to set the property "**SQL Index**" to *Document No., Document Type, Table ID, Line No., Dimension Code* for the PK and set the flag "**Clustered**". For older versions it is not recommended to change the PK in C/SIDE, here the index just should be changed on SQL Server site using TSQL:

```
USE [Navision]
GO

DROP INDEX [CRONUS$Document Dimension].[$1]

ALTER  TABLE [dbo].[CRONUS$Document Dimension]
DROP CONSTRAINT [CRONUS$Document Dimension$0]

ALTER  TABLE [dbo].[CRONUS$Document Dimension] WITH NOCHECK
ADD CONSTRAINT [CRONUS$Document Dimension$0]
PRIMARY KEY CLUSTERED
([Document No_],[Document Type],[Table ID],[Line No_],
[Dimension Code])

CREATE NONCLUSTERED INDEX [$1]
ON [dbo].[CRONUS $Document Dimension]
([Dimension Code],[Dimension Value Code])
```

>> Remove UNIQUE Flag and added PK Fields

Since NAV 4.0 (and higher) is it possible to remove the UNIQUE flag and additional PK fields by simply copying the content from "**Key**" to the property "**SQL Index**" (aka "*Basic Streamlining*").

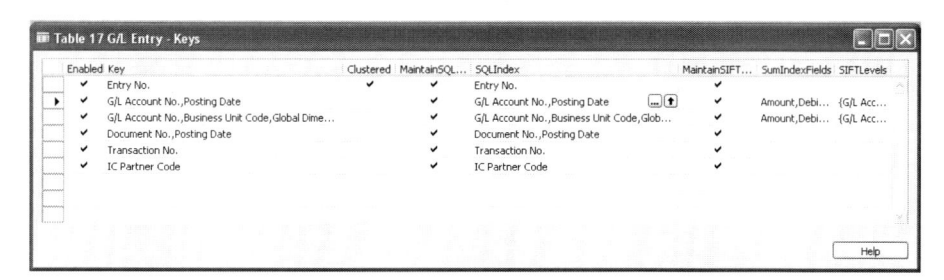

[Screenshot: Table 17 G/L Entry - Keys, showing Enabled, Key, Clustered, MaintainSQL, SQLIndex, MaintainSIFT, SumIndexFields, SIFTLevels columns with entries including Entry No.; G/L Account No., Posting Date; G/L Account No., Business Unit Code, Global Dime...; Document No., Posting Date; Transaction No.; IC Partner Code]

It was explained before that the "SQL Index" property should be handled with care. When copying from "*Key*" to "*SQL Index*" **1:1** the sorting of the index is not changed, it still matches to the "Key" sorting - hence, the problems described above should not occur!

So "*Basic Streamlining*" is not a full index optimization (regarding "*selectivity*" etc.), but a quick & simple – and 99% failsafe – way to improve indexes.

This should be only applied on large tables, with lots of records and/or many indexes.

STRYK System Improvement
Performance Optimization & Troubleshooting

With earlier version of NAV it is only possible to remove the UNIQUE flag and PK fields via TSQL:

Example:

Optimizing table 18 "*Customer*":
SK: *Search Name*

Standard NCI creation (UNIQUE with PK field):

```
USE [Navision]
GO

CREATE UNIQUE NONCLUSTERED INDEX [$1] ON
[dbo].[CRONUS$Customer]
([Search Name], [No_])
WITH DROP_EXISTING
```

Improved NCI creation:

```
USE [Navision]
GO

CREATE NONCLUSTERED INDEX [$1] ON [dbo].[CRONUS$Customer]
([Search Name])
WITH DROP_EXISTING
```

The major benefit here is that the indexes are getting much smaller, thus not wasting precious disk- and cache-space.

Caution: Do not remove the PK fields without removing the UNIQUE flag – this would cause severe "damage" to the business logic!

>> Define optimized Fill-Factors for the indexes

The **Fill-Factor** is the percentage up to which an index page is filled when rebuilt. This leaves some free space on the pages, so if data is inserted, SQL Server could write the information to this free space and does not have allocate a new index-page (which is costly!).

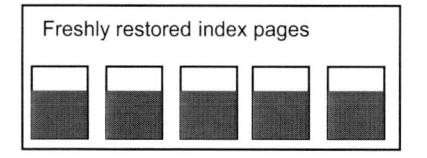

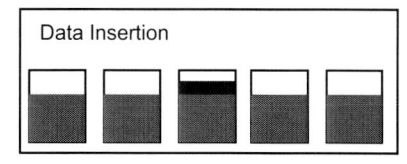

When an index page is completely filled, SQL Server allocates a new page and moves approximately the half amount of data to the "new" page to grant about 50% free space on the "old" page:

Original page | Data Insertion 100% occupied | Allocation of new page | Move 50% data to new page continue writing | Original page filled with ~50%

This operation, the allocation of new pages and data movement is called "**Page Split**". The less "*Page Splits/sec*" (Performance Monitor!) are performed, the faster transactions could be processed!

(The standard Fill-Factor in SQL Server is 0, which equals 100%; thus "out-of-the-box" pretty much "*Page Splits*" have to be performed)

Hence, if the Fill-Factor is **too high** – not enough free space on the pages – SQL Server has to perform many Page Splits. But if the Fill-Factor is **too low** – too much free space – then too much precious disk- and especially cache-space is wasted, as pages are always completely loaded!

STRYK System Improvement
Performance Optimization & Troubleshooting

For example, if a FF of 10% is defined, only 10% of the index contains data, the other 90% are just empty space, wasting precious cache space! On the one hand, this would speed up write transactions, of course, as just few Pages Splits have to be performed.

But the index would growth <u>tremendously</u>: with a FF of 10% there are 10-times more pages required to save the information, compared to a FF of 100%!

This would cause poor read-performance, resulting in high disk I/O slowing down the system.

Hence, it is crucial to define an <u>optimal</u> Fill-Factor for an index!

So, how to determine the optimal Fill-Factor?

Basically the Fill-Factor depends on the **growth** of data within a defined **period** (the *maintenance interval* in which the indexes are rebuilt).

As a rule of thumb one could say:

> **Fill-Factor = 100 – Growth Rate %**

So, if a table/index grows for e.g. 5%, the Fill-Factor could be 95%.

The NAV/SQL Performance Field Guide
Version 2009

The growth of data and indexes could be determined ideally with using the "**NAV Database Sizing Tool**", part of the "*NAV SQL Server Option Resource Kit*". Here, actually a stored procedure is used:

```
EXEC sp_spaceused [[@objname =] 'objname']
                  [,[@updateusage =] 'updateusage']
```

The size per index could be measured by using …

```
DBCC SHOWCONTIG
[ (
    { 'table_name' | table_id | 'view_name' | view_id }
    [ , 'index_name' | index_id ]
) ]
    [ WITH
        {
          [ , [ ALL_INDEXES ] ]
          [ , [ TABLERESULTS ] ]
          [ , [ FAST ] ]
          [ , [ ALL_LEVELS ] ]
          [ NO_INFOMSGS ]
        }
    ]
```
(SQL 2000/2005/2008)

… or querying …

```
sys.dm_db_index_physical_stats (
      { database_id | NULL | 0 | DEFAULT }
    , { object_id | NULL | 0 | DEFAULT }
    , { index_id | NULL | 0 | -1 | DEFAULT }
    , { partition_number | NULL | 0 | DEFAULT }
    , { mode | NULL | DEFAULT }
)
```

(SQL 2005/2008)

As mentioned above, the basic idea of Fill-Factors is to leave some reserved space to "squeeze in" data without allocating new pages. But for some indexes it is not necessary to have this free space, as data is never "squeezed in"!
For example, the **PK/CI** "Entry No." in several ledger entry tables: this is a field of type integer, and the records are *physically ordered* by this number (1, 2, 3, 4, …). There will never be data inserted between existing records, data is always added at the end – at the **last page**! Hence, the optimal Fill-Factor is 100%!

This could also be feasible for other PK/CI which are depending on "*No. Series*", but has to be handled with care (as "*No. Series*" could be changed).

Independent from the real table growth rate, a Fill-Factor should never be set to less than 70%, means the maximum free space granted should be 30%!

Setting the Fill-Factor via TSQL (example table 17 "G/L Entry"):

```
USE [Navision]
GO

CREATE UNIQUE CLUSTERED INDEX [CRONUS$G_L Entry$0]
ON [dbo].[CRONUS$G_L Entry]
([Entry No_])
WITH FILLFACTOR = 100,
DROP_EXISTING


CREATE NONCLUSTERED INDEX [$1]
ON [dbo].[CRONUS$G_L Entry]
([G_L Account No_],[Posting Date])
WITH FILLFACTOR = 90,
DROP_EXISTING
```

If a growth-based Fill-Factor cannot be applied, it could be feasible to implement a FF of 90 to 95%; e.g. using a "**Maintenance Plan**" (*Caution: Within MP not the FF is defined but the free space; hence, to apply a FF of 90% the "Free Space Amount" has to be defined as 10%*)

Even with setting the optimal Fill-Factors, the structure of an index will fragment over time. There are different types of fragmentation on Page or Extent level.

The degree of fragmentation could be viewed using the **DBCC SHOWCONTIG** statement (2000/2005/2008) or **sys.dm_db_index_physical_stats** (2005/2008).

DBCC SHOWCONTIG:

Output	Best Value	Explanation
Pages Scanned	n/a	Number of pages in the table or index.
Extents Scanned	n/a	Number of extents in the table or index.
Extent Switches	= Extents Scanned - 1	Number of times the DBCC statement moved from one extent to another while the statement traversed the pages of the table or index.
Average Pages Per Extent	n/a	Number of pages per extent in the page chain.
Scan Density [Best:Actual]	>= 90	Is a percentage. It is the ratio Best Count to Actual Count. This value is 100 if everything is contiguous; if this value is less than 100, some fragmentation exists. Best Count is the ideal number of extent changes if everything is contiguously linked. Actual Count is the actual number of extent changes.
Logical Scan Fragmentation	<= 10	Percentage of out-of-order pages returned from scanning the leaf pages of an index. This number is not relevant to heaps. An out-of-order page is one for which the next page indicated in an IAM is a page different from the page pointed to by the next page pointer in the leaf page.
Extent Scan Fragmentation	<= 10	Percentage of out-of-order extents in scanning the leaf pages of an index. This number is not relevant to heaps. An out-of-order extent is one for which the extent that contains the current page for an index is not physically the next extent after the extent that contains the previous page for an index.
Average Bytes Free Per Page	n/a	Average number of free bytes on the pages scanned. The larger the number, the less full the pages are. Lower numbers are better if the index will not have many random inserts. This number is also affected by row size; a large row size can cause a larger number.
Average Page Density (Full)	>= 80	Average page density, as a percentage. This value takes into account row size. Therefore, the value is a more accurate indication of how full your pages are. The larger the percentage, the better.

STRYK System Improvement
Performance Optimization & Troubleshooting

To de-fragment the indexes and to restore the Original Fill-Factors it is mandatory to run periodic maintenance. This could be done by

- **DROPPING** the indexes and **CREATE** them from the scratch (best result, but time consuming and blocking) – not really recommended
- Using the **ALTER INDEX REBUILD** (2005/2008) or **DBCC DBREINDEX** (2000) feature (good result, less time consuming, but blocking – also executed by "Maintenance Plan")
- Using the **ALTER INDEX REORGANIZE** (2005/2008) or **DBCC INDEXDEFRAG** feature (2000) (moderate result, less blocking)

Basically the difference between *REORGANIZE/DEFRAG* and *REBUILD/REINDEX* is this:
The *REORGANIZE* just defragments the "**Leaf Nodes**" of an index
The *REBUILD* defragments the all "**Index Nodes**"

Recommended Index-Defragmentation-Scale:

Logical Scan Fragmentation < 10%   → No Index-Maintenance required
Logical Scan Fragmentation >= 10% & < 30% → REORGANIZE or DEFRAG
Logical Scan Fragmentation >= 30%   → REBUILD or DBREINDEX

Example - Rebuilding Index with FF of 90%:

```
DBCC DBREINDEX ('CRONUS$Sales Line', 'CRONUS$Sales Line$0', 90)
```

or

```
ALTER INDEX 'CRONUS$Sales Line$0' ON 'CRONUS$Sales Line'
REBUILD WITH (FILLFACTOR = 90)
```

(See chapter about "*Maintenance*" blow for an example job!)

>> Index Statistics

According to the general structure of indexes it is also most important to have sufficient **Index Statistics**. The **Query Optimizer** of the SQL Server "decides" on basis of these statistics which index to use for a specific query. The statistics contain information about the **selectivity** of the index at all and each index field. Missing, wrong or outdated statistics would cause that "bad" indexes are used; performance is decreased.

View existing statistics:

```
EXEC sp_helpstats [@objname = ] 'object_name'
                  [,[@results = ] 'value']
```

As mentioned previously, it is not recommended to maintain the statistics automatically. It is better to have a SQL Server Agent **job** running, which updates them daily.

TSQL for the SQL Server Agent job:

```
EXEC sp_updatestats
GO
EXEC sp_createstats 'indexonly'
```

Statistics which were created by the "*Auto. Create Statistics*" feature have to be erased (see above); only sufficient Index Statistics are required and should remain.

Caution:
User statistics which were created by "sp_createstats" cannot be automatically deleted by an `ALTER TABLE ALTER COLUMN` command. This means, that when changing e.g. the name or data-type of a field within a NAV table (using "*Object Designer*") an error would be raised.
If this problem occurs, the user-statistics could be deleted. Then the required change could be performed, and afterwards the statistics can be recreated by the `sp_createstats 'indexonly'` command.

STRYK System Improvement
Performance Optimization & Troubleshooting

Dropping User-Statistics:

```
DECLARE @id INT, @name VARCHAR(128), @statement NVARCHAR(1000)
DECLARE stat_cur CURSOR FAST_FORWARD FOR
  SELECT [id], [name] FROM sysindexes
  WHERE (indexproperty([id], [name], N'IsStatistics') = 1)
    AND (indexproperty([id], [name], N'IsAutoStatistics') = 0)
    AND (isnull(objectproperty([id], N'IsUserTable'),0) = 1)
  ORDER BY object_name([id])
OPEN stat_cur
FETCH NEXT FROM stat_cur INTO @id, @name
  WHILE @@fetch_status = 0 BEGIN
    SET @statement = 'DROP STATISTICS [' +
                     object_name(@id) + '].[' + @name + ']'
    BEGIN TRANSACTION
      PRINT @statement
      EXEC sp_executesql @statement
    COMMIT TRANSACTION
    FETCH NEXT FROM stat_cur INTO @id, @name
  END
CLOSE stat_cur
DEALLOCATE stat_cur
```

>> Move indexes to a dedicated File-Group

With SQL Server it is possible to store data in different files and **File-Groups**. Here the data could be physically stored on different devices to benefit from a <u>dedicated disk I/O</u>.

Remark:
Previously it was "assumed", that File/ File-Group is handled with a separate **Thread** on SQL Server, thus the parallelism might be increased to further improve performance. Well, there are articles (www) which explain this might be a **fairy tale** …
Hence, the only reason for splitting up a database should be to store each file on a separate disk!
See http://blogs.msdn.com/psssql/archive/2007/02/21/sql-server-urban-legends-discussed.aspx

There are a lot of strategies about how to distribute data on different files/file-groups. What could be feasible is, to store the **Non-Clustered Indexes** on such a dedicated File-Group!

Creating a new File-Group - Example:

```
USE [master]
GO

ALTER DATABASE [Navision] ADD FILEGROUP [IndexGroup]
GO

ALTER DATABASE [Navision] ADD FILE (
  NAME = N'Navision_Idx',
  FILENAME = N'D:\Databases\Navision_Idx.ndf',
  SIZE = 10240000KB,
  FILEGROWTH = 102400KB)
TO FILEGROUP [IndexGroup]
GO
```

The initial size for the new "*IndexGroup*" should be about **30%** of the data-size (usually "*Data Filegroup 1*")

TSQL to move indexes into a File-Group (example table 17 "G/L Entry"):

```
USE [Navision]
GO

CREATE UNIQUE CLUSTERED INDEX [CRONUS$G_L Entry$0] ON
[dbo].[CRONUS$G_L Entry]
([Entry No_])
WITH FILLFACTOR = 100,
DROP_EXISTING
GO

CREATE NONCLUSTERED INDEX [$1]
ON [dbo].[CRONUS$G_L Entry]
([G_L Account No_],[Posting Date])
WITH FILLFACTOR = 90,
DROP_EXISTING
ON IndexGroup
GO
```

Caution: Do _not_ move the Clustered Indexes somewhere else. The "leaf node" of the CI is the table – do not separate the table from this index!

>> Specific Hints for Index Creation in SQL Server 2005/2008

With SQL Server 2005/2008 some advanced hints could be set when creating indexes.

Example:

```
CREATE NONCLUSTERED INDEX [$1] ON [dbo].[CRONUS$G_L Entry]
(
        [G_L Account No_] ASC,
        [Posting Date] ASC
)
WITH (PAD_INDEX = OFF,
      SORT_IN_TEMPDB = OFF,
      DROP_EXISTING = ON,
      IGNORE_DUP_KEY = OFF,
      ONLINE = OFF,
      MAXDOP = 64,
      FILLFACTOR = 90)
ON [IndexGroup]
```

Here especially the **ONLINE** feature (Enterprise Edition) is to be mentioned: If set to ON the SQL Server will not hold long-term locks during the index operation. This is slower, but could avoid blocks.

When the "Degree of Parallelism" was changed (see chapter "Configuration") then it is recommended to set the **MAXDOP** parameter to 64 (maximum) to use all CPU for parallel indexing.

If tables are heavily used with for data insertion, it could be feasible to set the **PAD_INDEX** parameter to ON – in this case a **FILLFACTOR** will be applied to all "*Index Nodes*" (else only "*Leaf Nodes*" get a FF).
This would generally increase the size of an index, but could speed up write-transactions.

STRYK System Improvement
Performance Optimization & Troubleshooting

> Structure of SIFT Indexes in NAV

SIFT = **Sum Index Flowfield® Technology**

This patented feature was actually invented by Navision. It allows seeing **aggregated** values (e.g. sums) within a record without physically storing them in the table. SQL Server does not support this, so it is somewhat "*emulated*":

If a **Flowfield** should be created in C/SIDE it is necessary to have a key/index (SIFT Index) which contains all fields which are used for the aggregation and the to-be-aggregated field as "**SumIndexField**" on the detail-table.
If this is done with SQL Server option, C/SIDE generates a specific table (SIFT Table) for each of the SIFT Indexes where the aggregated values will be stored.

Example, table 32 "Item Ledger Entry":

The first SIFT Index here is the second key:

Fields: Item No., Variant Code, Drop Shipment, Location Code, Posting Date
SumIndexFields: Quantity, Invoiced Quantity

The SIFT Table creates on SQL Server site is "CRONUS$32$0".

Its columns are *bucket, f2, f5402, f47, f8, f3, s12, s14*.

Here the f-fields refer to the field-IDs of the SIFT Index fields ($f2$ = *Item No.*, $f5402$ = *Variant Code*, etc.), the s-fields to the SumIndexFields ($s12$ = *Quantity*, $s14$ = *Invoiced Quantity*).

The *bucket* defines the level of aggregation. These levels can be enabled in C/SIDE (**SIFTLevelsToMaintain**):

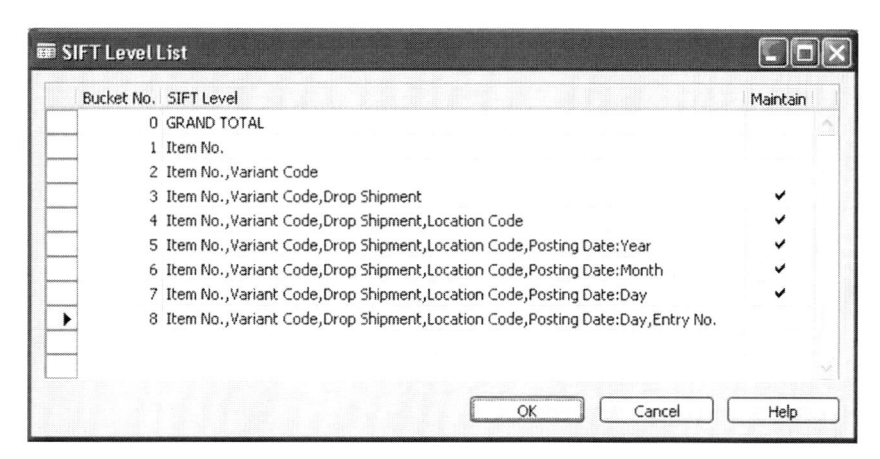

Bucket No.	SIFT Level	Maintain
0	GRAND TOTAL	
1	Item No.	
2	Item No.,Variant Code	
3	Item No.,Variant Code,Drop Shipment	✓
4	Item No.,Variant Code,Drop Shipment,Location Code	✓
5	Item No.,Variant Code,Drop Shipment,Location Code,Posting Date:Year	✓
6	Item No.,Variant Code,Drop Shipment,Location Code,Posting Date:Month	✓
7	Item No.,Variant Code,Drop Shipment,Location Code,Posting Date:Day	✓
8	Item No.,Variant Code,Drop Shipment,Location Code,Posting Date:Day,Entry No.	

Now, whenever a record is inserted into the (detail-)table, a SQL Server site **Trigger** is processed to fill/update the SIFT Table. Means, the **SQL trigger** is summing up the values on different aggregation levels – buckets – and saves the summary records in the SIFT table.

The more SIFT buckets have to be filled (**Costs per Record**), and the more records are save in the detail-table, the longer is the duration for updating the SIFT records, thus slowing down the system.

To improve performance, the number of SIFT levels/buckets has to be reduced. In tables which contain a reasonable number of records and have a high dynamic on insertion and deletion ("*hot*" tables), like "Sales Line", or "Purchase Line" , basically all SIFT levels could be deleted (**MaintainSIFTIndex** = FALSE).

For larger tables it is feasible, to enable just the **most detailed bucket** (except the last ones, which include the PK fields) as all lower buckets could be generated from this.

STRYK System Improvement
Performance Optimization & Troubleshooting

In our example above, only bucket *number 7* should be enabled, for all other "**Maintain**" should be set to FALSE.

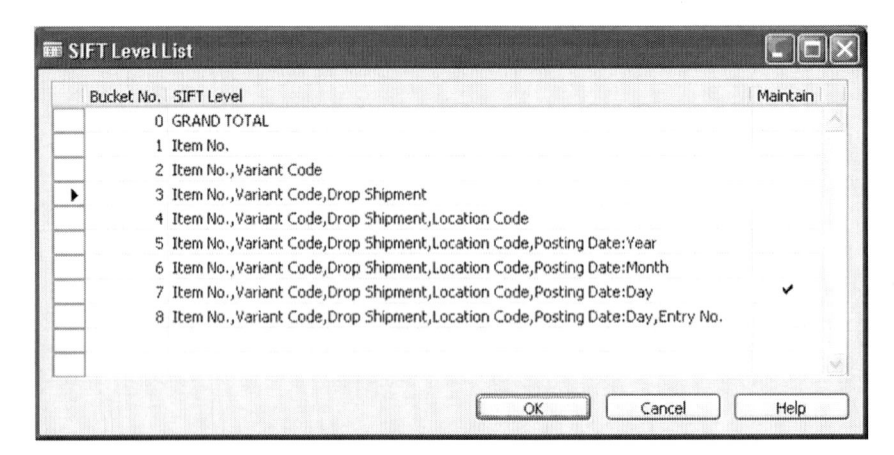

```
SIFT Level List
Bucket No.  SIFT Level                                                              Maintain
         0  GRAND TOTAL
         1  Item No.
         2  Item No.,Variant Code
       ▶ 3  Item No.,Variant Code,Drop Shipment
         4  Item No.,Variant Code,Drop Shipment,Location Code
         5  Item No.,Variant Code,Drop Shipment,Location Code,Posting Date:Year
         6  Item No.,Variant Code,Drop Shipment,Location Code,Posting Date:Month
         7  Item No.,Variant Code,Drop Shipment,Location Code,Posting Date:Day      ✓
         8  Item No.,Variant Code,Drop Shipment,Location Code,Posting Date:Day,Entry No.

                                        OK          Cancel          Help
```

This reduces the "**Costs per Record**" remarkably, speeding up *write* transactions! Further, to improve the queries on the SIFT tables (from the SQL Trigger) in some cases it could be feasible to create a **covering index** (includes all fields of the table) there, like

```
CREATE INDEX ssi_CovIdx ON dbo.[CRONUS$32$0]
(bucket, f2, f5402, f47, f8, f3, s12, s14)
```

Here the *reading* speed is increased!

During the frequent insertion and update of records in the SIFT tables, there will be records where all sum-fields equal zero, which are not deleted automatically. For C/SIDE there is no difference if a query on a SIFT table returns a record where all sums are zero or if no record is retrieved at all; or for example:
1 + 0 + 2 is the same like 1 + 2; the result is the same (3).

Hence, these records – **Empty SIFT Record** – should be deleted to reduce the number of records in the table.

Example:

```
DELETE FROM [CRONUS$32$0] WHERE (s12 = 0) and (s14 = 0)
```

Caution 1: It is absolutely crucial to only delete those records where **all** sum-fields contain the value zero, otherwise the deletion would cause severe "damage" to the business logic!

Caution 2: Some NAV versions have a bug within the SIFT management: When performing this kind of SIFT maintenance – regardless if by "Table Optimizer" or other procedures – FlowFields may display wrong values. Updates/Hotfixes were provided to solve this problem!

Affected database-versions are (as far as known – errors excepted):

- NAV 4.0:   all versions before 63
- NAV 5.0:   all versions before 82

(See Appendix B for details)

And have in mind that since NAV 5.0 SP1 the SIFT tables were replaced by VSIFT Views; hence no maintenance is required at all!

STRYK System Improvement
Performance Optimization & Troubleshooting

>> Moving SIFT to a dedicated File-Group

For the same reason as for moving NCI to a dedicated file-group, it could be feasible to store the SIFT tables (maybe including indexes) in a <u>separate</u> **file-group** as well.

With SQL Server 2000 this requires to 1) rename the original SIFT table, 2) re-create the original SIFT table in new file-group, 3) copy all data from renamed to new SIFT table and 4) re-create all constraints and indexes on the new SIFT table.

Creating new File-Group - Example:

```
USE [master]
GO

ALTER DATABASE [Navision] ADD FILEGROUP [SIFTGroup]
GO

ALTER DATABASE [Navision] ADD FILE (
  NAME = N'Navision_SIFT',
  FILENAME = N'D:\Databases\Navision_SIFT.ndf',
  SIZE = 5120000KB,
  FILEGROWTH = 102400KB)
TO FILEGROUP [SIFTGroup]
GO
```

Since SQL Server 2005 the **MOVE TO** clause could be used - Example:

```
USE [Navision]
GO

BEGIN TRANSACTION

ALTER TABLE dbo.[CRONUS$17$0]
DROP CONSTRAINT [CRONUS$17$0_idx]
WITH (MOVE TO [SIFTGroup])

ALTER TABLE dbo.[CRONUS$17$0] WITH NOCHECK
ADD CONSTRAINT [CRONUS$17$0_idx]
PRIMARY KEY CLUSTERED ([bucket],[f3],[f4])
ON [SIFTGroup]

CREATE INDEX [CRONUS$17$0_hlp_idx] ON dbo.[CRONUS$17$0] ([f4])
WITH (DROP_EXISTING = ON) ON [SIFTGroup]

COMMIT TRANSACTION
GO
```

> Structure of VSIFT Views in NAV

With NAV Version 5.0 **Service Pack 1** the SIFT Indexes (with its SIFT tables) have been replaced by so called "**Indexed Views**" aka **VSIFT**

The SQL Server object "View" is actually a pre-defined SELECT statement querying the data from a table. While these VSIFT Views are querying the data e.g. from the "Ledger Entry" tables, the defined "**SumIndexFields**" are summed up grouped by the defined "**Key**" fields.

Example:

Table:	17 G/L Entry
Key:	G/L Account No., Posting Date
SumIndexFields:	Amount, Debit Amount, Credit Amount, Additional-Currency Amount, Add.-Currency Debit Amount, Add.-Currency Credit Amount

VSIFT:

```
CREATE VIEW [dbo].[CRONUS$G_L Entry$VSIFT$1]
WITH SCHEMABINDING
AS
SELECT "G_L Account No_",
       "Posting Date",
       COUNT_BIG(*) "$Cnt",
       SUM("Amount") "SUM$Amount",
       SUM("Debit Amount") "SUM$Debit Amount",
       SUM("Credit Amount") "SUM$Credit Amount",
       SUM("Additional-Currency Amount")
           "SUM$Additional-Currency Amount",
       SUM("Add_-Currency Debit Amount")
           "SUM$Add_-Currency Debit Amount",
       SUM("Add_-Currency Credit Amount")
           "SUM$Add_-Currency Credit Amount"
FROM dbo."CRONUS$G_L Entry"
GROUP BY "G_L Account No_","Posting Date"
```

Here the naming is more transparent than with the SIFT Tables, thus the VSIFT could be related to the relevant "Key" easier.

Comparing this VSIFT with a similar SIFT structure it could be seen, that the VSIFT aggregation level more or less <u>equals</u> (!) to the *pre-last* SIFT bucket (see above), thus aggregating one level "above" Primary Key level.

The `SCHEMABINDING` allows to create Indexes on this "View"! To do this on aggregating views it is also necessary to include the `COUNT_BIG` field (see *"SQL Server Books Online"* for details).

By default a **Clustered Index** is generated for each VSIFT View:

```
CREATE UNIQUE CLUSTERED INDEX [VSIFTIDX]
ON [dbo].[CRONUS$G_L Entry$VSIFT$1]
(
        [G_L Account No_] ASC,
        [Posting Date] ASC
)
```

This index supports the reading from the source table, especially when filters on the VSIFT are applied.

<u>Caution</u>: With some older NAV builds, the VSIFT of a table are always re-created if the table is *recompiled* within Object Designer, even though no changes to the Keys or Indexes were made. This includes the re-creation of the Clustered Indexes of the VSIFT which could be somewhat time consuming on large tables!

>> Comparing SIFT and VSIFT

With VSIFT it is not necessary to write additional **data** to tables as it is done with SIFT, hence with VSIFT no additional "*Cost Per Record*" are caused, the writing of data happens as fast as possible.
Of course, as no SIFT records are created, the database gets **much** smaller which has other advantages, too (e.g. faster & smaller backups etc.).

With standard NAV too many SIFT buckets are enabled, causing too much "*Cost Per Record*", slowing down read- and write-performance.

So, of course, there is a remarkable difference in *writing*-performance when comparing standard SIFT with standard VSIFT – VSIFT performs much faster!

But after **optimizing** the SIFT structure – reducing the number of "*buckets*" – the difference in write-performance is not that big anymore:

An optimized SIFT structure provides the sum-data on the same level as with VISIFT!

PRO – SIFT:
- ✓ Zero-Sums can be deleted (= less records = faster reads)
- ✓ Number of Buckets could be increased on demand (faster reading on higher aggregation levels)
- ✓ Easier to optimize

CONTRA – SIFT:
- ✗ Slow Trigger execution
- ✗ More space consuming
- ✗ Higher blocking risk

PRO – VSIFT:
- ✓ Less space consuming
- ✓ Faster update (no trigger code)
- ✓ Less blocking issues

CONTRA – VSIFT:
- ✗ Zero-Sums cannot be deleted (= more records = slower reads)
- ✗ No higher aggregation levels ("buckets") available
- ✗ Difficult to optimize

This means the *advantages* of the one are the *disadvantages* of the other one …

Hence, just the VSIFT feature alone should not necessarily be a reason for an upgrade; SIFT is not that bad (after tuning).

As mentioned, VSIFT are quite tricky to tune; doing this from NAV might require to add/change "Keys" and to change some C/AL programming.
If a higher aggregate is required, with SIFT it is possible to simple re-activate this bucket; with VSIFT another "Key" has to be created.

Example – Aggregation on "G/L Account No.":

Table:	17 G/L Entry
Key:	G/L Account No.
SumIndexFields:	Amount, Debit Amount, Credit Amount, Additional-Currency Amount, Add.-Currency Debit Amount, Add.-Currency Credit Amount

To minimize the additional "load" on the system, the "*MaintainSQLIndex*" property should be set to FALSE (as a similar index including "G/L Account No." already exists), the flag "*MaintainSIFTIndex*" has to be TRUE.

In this example a <u>new</u> VSIFT View would be generated, aggregation on "*G/L Account No.*" level only. According to this, additional changes within the C/AL Code may be required.

STRYK System Improvement
Performance Optimization & Troubleshooting

In many cases reading from VSIFT could be improved by applying a **Covering Index** or an Index with **Included Columns**:

Example – Covering Index:

```
CREATE NONCLUSTERED INDEX ssi_CovIdx
ON dbo.[CRONUS$G_L Entry$VSIFT$1]
(
   [G_L Account No_],
   [Posting Date],
   [SUM$Amount],
   [SUM$Debit Amount],
   [SUM$Credit Amount],
   [SUM$Additional-Currency Amount],
   [SUM$Add_-Currency Debit Amount],
   [SUM$Add_-Currency Credit Amount]
)
```

Example – Index Included Columns:

```
CREATE NONCLUSTERED INDEX ssi_CovIdx
ON dbo.[CRONUS$G_L Entry$VSIFT$1]
(
   [G_L Account No_]
)
INCLUDE
(
   [SUM$Amount]
)
```

(Caution: Recompiling the object might delete these indexes with older NAV versions!)

The NAV/SQL Performance Field Guide
Version 2009

>> Replacing SIFT/VSIFT using Included Columns

In some cases – if the table contains not too many records; e.g. less than 10.000 – it might be feasible to disable all SIFT/VSIFT Indexes anyway.

In this case it could still happen, that calculating sums from it perform sub-optimal, causing too many Reads. Instead of enabling a SIFT or VSIFT an Index with **Included Columns** could be created:

Example:

Table: 37 Sales Line
Key: Document Type, Type, No., Variant Code, Drop Shipment, Location Code, Shipment Date
SumIndexFields: Outstanding Qty. (Base)
MaintainSQLIndex: **FALSE**
MaintainSIFTIndex: **FALSE**

Compensation with NCI (INCLUDE):

```
CREATE NONCLUSTERED INDEX ssi_IdxInc
ON dbo.[CRONUS$Sales Line]
(
        [Document Type], [Type], [No_], [Variant Code],
        [Drop Shipment], [Location Code], [Shipment Date]
)
INCLUDE
(
        [Outstanding Qty_ (Base)]
)
```

> Design of C/AL Code

The general problem with the C/AL code in NAV is, that it was written for the "native" database server. There is "ISAM" (*Index Sequential Access Method*) and "*Version Principle*" on FDB site versus "*Relational Database*" and "*Result Set*" on SQL site.

Besides the **technical options** to improve the source code, it is even more important to **design processes** optimal: minimizing the number of queries, properly sizing result-sets and ideal definition of the transaction lengths.

The general process design is off scope of this document as this has to be investigated and optimized **individually**.

The following gives some technical advices, regarding the programming with C/AL.

The NAV/SQL Performance Field Guide
Version 2009

>> Querying SQL Server

It is necessary to understand, how queries are sent from C/SIDE to the SQL Server, as the C/AL code has to be translated into SQL:

C/AL	SQL
Record.SETCURRENTKEY(Field1, Field2) Record.SETRANGE(Field1, Value1) Record.SETFILTER(Field2, Value2) IF Record.FIND('-') THEN …	SELECT * FROM Company$Record WHERE (Field1 = Value1) AND (Field2 = Value2) ORDER BY (Field1, Field2, PK)

The **SETCURRENTKEY** statement only defines the sorting, the **ORDER BY** clause in SQL – it's not determining which index is used for retrieval!
SETRANGE and **SETFILTER** define the **WHERE** clause; here both statement have the same "value" (in C/SIDE **SETRANGE** performs different than **SETFILTER**).
The **FIND** statement empties the C/SIDE command buffer and sends the query.
While in C/SIDE it is important to set the filters in order of the used key fields, for SQL Server this does not matter.

Additionally, C/SIDE sets some optimizer hints as **TOP**, **READUNCOMMITTED**, **UPDLOCK**, **OPTION FAST** or other things..

The <u>simpler</u> the SQL statement is created, the faster it could be executed!

To create simple statements, some rules should be followed:

× Only use **SETCURRENTKEY** if the sorting is important or to avoid problems with "SQL Cursor" declaration
× Avoid **multiple filtering** on the same field (this is caused by the combination of **TableView** properties, **FILTERGROUPS** and C/AL)
× Avoid usage of "**wildcards**" (* or ?) when filtering, especially at the beginning of an expression

STRYK System Improvement
Performance Optimization & Troubleshooting

C/AL Code to SQL Translation (Examples):
(NAV 4.00 and higher)

C/AL	SQL (combined)	CSR
Customer.SETRANGE ("Country Code", 'US'); Customer.FIND('-');	SELECT *,DATALENGTH("Picture") FROM "dbo"."Cronus$Customer" WITH (READUNCOMMITTED) WHERE (("Country Code"='US')) ORDER BY "No_"	!
Customer.SETRANGE ("Country Code", 'US'); Customer.FINDFIRST;	SELECT **TOP 1** *,DATALENGTH("Picture") FROM "dbo"."Cronus$Customer" WITH (READUNCOMMITTED) WHERE (("Country Code"='US')) ORDER BY "No_"	
Customer.SETRANGE ("Country Code", 'US'); Customer.FIND('+');	SELECT *,DATALENGTH("Picture") FROM "dbo"."Cronus$Customer" WITH (READUNCOMMITTED) WHERE (("Country Code"='US')) ORDER BY "No_" **DESC**	!
Customer.SETRANGE ("Country Code", 'US'); Customer.FINDLAST;	SELECT **TOP 1** *,DATALENGTH("Picture") FROM "dbo"."Cronus$Customer" WITH (READUNCOMMITTED) WHERE (("Country Code"='US')) ORDER BY "No_" **DESC**	
Customer.SETRANGE ("Country Code ", 'US'); Customer.FINDSET;	SELECT **TOP 500** *,DATALENGTH("Picture") FROM "dbo"."Cronus$Customer" WITH (READUNCOMMITTED) WHERE (("Country Code"='US')) ORDER BY "No_"	!
Customer.SETRANGE ("Country Code ", 'US'); Customer.FINDSET(TRUE);	**SET TRANSACTION ISOLATION LEVEL SERIALIZABLE** SELECT *,DATALENGTH("Picture") FROM "dbo"."Cronus$Customer" WITH (**UPDLOCK**) WHERE (("Country Code"='US')) ORDER BY "No_"	!

CSR = Cursor

C/AL	SQL (combined)	CSR
Customer.SETRANGE ("Country Code", 'US'); IF Customer.ISEMPTY THEN;	SELECT **TOP 1 NULL** FROM "dbo"."Cronus$Customer" WITH (READUNCOMMITTED) WHERE (("Country Code"='US'))	
Customer.SETRANGE ("Country Code", 'US'); IF Customer.COUNT <> 0 THEN;	SELECT **COUNT**(*) FROM "dbo"."Cronus$Customer" WITH (READUNCOMMITTED) WHERE (("Country Code"='US'))	
Customer.SETRANGE ("Country Code", 'US'); IF Customer.COUNTAPPROX <> 0 THEN;	**SET SHOWPLAN_ALL ON** SELECT * FROM "dbo"."Cronus$Customer" WITH (READUNCOMMITTED) WHERE (("Country Code"='US'))	
GLEntry.LOCKTABLE; GLEntry.FIND('+');	**SET TRANSACTION ISOLATION LEVEL SERIALIZABLE** SELECT * FROM "dbo"."Cronus$G_L Entry" WITH (**UPDLOCK**) ORDER BY "Entry No_" **DESC**	!
GLEntry.LOCKTABLE; GLEntry.FINDLAST;	**SET TRANSACTION ISOLATION LEVEL SERIALIZABLE** SELECT **TOP 1** * FROM "dbo"."Cronus$G_L Entry" WITH (**UPDLOCK**) ORDER BY "Entry No_" **DESC**	

CSR = Cursor

>> Cursor Handling

With C/SIDE most queries are performed as so called "**cursor** operations". A cursor is, simplified, a result-set of record which is browsed (cursor fetch) record by record for further processing (similar to a loop within NAV).
Unfortunately, cursor handling is "costly", as each cursor has to be maintained in the SQL Servers cache, thus reducing performance.

A cursor is necessary when processing a result-set within a loop (**REPEAT** ... **UNTIL**, **WHILE** ... **DO**).
Usually a loop starts with a `FIND('-')` (or `FINDSET`), but this statement is also often used just to check if a filtered range is empty, or just to process the first record. But C/SIDE is not able to distinguish this, thus it is always generating cursors!

With NAV 4.00 there were C/AL command introduced, which could reduce the number of cursor creations.

- ✓ If processing a loop, use **FINDSET** instead of `FIND('-')`
  When modifying data – the result-set - within the loop use **FINDSET(TRUE)**
  When modifying a field which is part of the currently used Key use
  FINDSET(TRUE, TRUE)
  (`FINDSET` cannot be used with a *descending* order)
- ✓ If only processing the first record of a set, use **FINDFIRST** instead of `FIND('-')`
- ✓ If only processing the last record of a set, use **FINDLAST** instead of `FIND('+')`
- ✓ If checking if a result is empty, use **ISEMPTY**

Especially the design of loops has severe impact on the cursor handling and performance:

- ✗ Never use something like `REPEAT ... UNTIL NOT FIND('-')`
- ✗ Never change the instance of a record variable while processing itself, use a *browsing instance* and a *working instance*.

Wrong	Correct

`Rec.SETRANGE(Field1, Value1);`
`IF Rec.FIND('-') THEN`
`  REPEAT`
`    Rec.Field1 := Value2;`
`    Rec.MODIFY;`
`  UNTIL NOT Rec.FIND('-');`
	`Rec1.SETRANGE(Field1, Value1);`
	`IF Rec1.FINDSET(FALSE, FALSE) THEN`
	`  REPEAT`
	`    Rec2.GET(Rec1.PrimaryKey);`
	`    Rec2.Field1 := Value2;`
	`    Rec2.MODIFY;`
	`  UNTIL Rec1.NEXT = 0;`

The NAV/SQL Performance Field Guide
Version 2009

>> Using SQL Server code

One major advantage of SQL Server is, that it basically is possible to perform **server-site processing**, with FDB not. Unfortunately, C/SIDE is not using this feature, all business logic is processed at the NAV Client.

Nonetheless, SQL Server could do a lot of processes much (!!!) faster than the NAV Client! To benefit from this, it is sometimes feasible to "leave" C/SIDE and execute SQL directly.

The easiest way to execute SQL is by using the command-line utility **OSQL, ISQLW** (the **Query Analyzer** (SQL 2000)) or **SQLCMD** (SQL 2005/2008) - Examples:

```
ReturnValue :=
   SHELL(STRSUBSTNO('isqlw -E -S "%1" -i "%2" -o "%3"',
                    <SQL_Server_Name>,
                    <TSQL_Script_File>,
                    <Output_File>));
```

or

```
ReturnValue :=
   SHELL(STRSUBSTNO(' sqlcmd -S "%1" -d "%2" -E -Q "%3" -o "%4"',
                    <SQL_Server_Name>,
                    <Database>,
                    <TSQL_Query>,
                    <Output_File>));
```

(In these cases a **trusted connection** (–E) (*Windows Authentication*) is used.)

Even more convenient is the usage of **MS ADO** (*ActiveX Data Object*) as **Automation Server**: A <u>direct</u> connection to the SQL Server could be established through the OLE-DB provider **SQLOLEDB** and queries can be sent directly and the received result-set could be processed directly in NAV:

```
IF ISCLEAR(ADOConnection) THEN
  CREATE(ADOConnection);
ConnectionString := 'Server=' + <Server_Name>;

IF <Trusted> THEN  // Windows Authentication
    ConnectionString += ';Trusted_Connection=Yes;'
ELSE  // Database Authentication
  ConnectionString +=
  ';Trusted_Connection=no;UID=' + <User_ID> + ';pwd='
  + <Passwd> + ';'

CursorLocation := 3;  // Client Site

ADOConnection.Provider('SQLOLEDB');
ADOConnection.CursorLocation(CursorLocation);
ADOConnection.ConnectionString(ConnectionString);
ADOConnection.Open;

RecordsAffected := 0;
Options := 1;  // Textual Definition

IF ADOConnection.State = 1 THEN BEGIN  // connected
  ADOConnection.DefaultDatabase(<Database_Name>);
  ADOConnection.CommandTimeout(0);  // Infinite
  ADORecSet :=
    ADOConnection.Execute(<Statement>, RecordsAffected,
                          Options);
  ADOErrors := ADOConnection.Errors();
END;

ADOConnection.Close;
```

It has to be mentioned, that the length of the **<Statement>** is limited to the NAV restrictions: A text variable could be defined with just 1024 characters. If longer statements have to be sent an **array** of this variable could be used and sent like

```
ADORecSet :=
  ADOConnection.Execute(Statement[1] + Statement[2],
  RecordsAffected, Options);
```

(If this is not sufficient or inconvenient, then the OSQL/ISQLW/SQLCMD option should be preferred.)

Caution: If this ADO connection is used frequently, it is not recommended to always "*connect-execute-disconnect*", as the login procedure is time-consuming. Here it is better to e.g. *connect* with the **first** usage, then leave the connection *established* and *disconnect* when the NAV client is **closing**.

STRYK System Improvement
Performance Optimization & Troubleshooting

>> Linked Objects

In NAV it is also possible to integrate SQL Server site "Views" as so called "**Linked Object**" (Table Property, see "*C/SIDE Reference Guide*" for details).
Such a "View" could combine data from multiple NAV tables, or data from other databases on the same Server – or from *other* Servers or data-sources, attached as "**Linked Server**" in SQL Server (see "*Books Online*" for details).

For example, this would make it possible to establish a "Linked Server" on a Oracle System, then create a "View" on specific Oracle-DB-Tables (e.g. historical data) and then use this "View" as "Linked Object" in NAV ("Linked Objects" can be used as any other NAV table!).
Hence, the historical data from the Oracle system could be used within the NAV business logic – like any other NAV table!

So, with SQL Server there are a LOT possibilities to **integrate** with other systems (data exchange, data migration, etc.), one of the major advantages to the C/SIDE Server.

But have in mind that using "Linked Objects" may perform not really fast, (partly) depending on the data-source which is used.

Also, when creating such "Views" it is important to only use SQL Data-Types which are compatible with the NAV Data-Types (this is <u>crucial</u> when writing data from external applications/programs into NAV tables):

NAV Type	SQL Type	Example	Comment
Integer	int	2.147.483.647	
BigInteger	bigint	455.500.000.000	
Decimal	decimal	3.425,57	
Text	varchar	Hello World	
Code	varchar	HELLO	uppercase only
Option	int	1	
Boolean	tinyint	0	
Date	datetime	2008-09-24 **00:00:00.000**	
Time	datetime	**1754-01-01** 12:53:00.000	
DateTime	datetime	2008-09-24 10:53:00.000	UTC shift
Binary	varbinary		
BLOB	image		compressed

DateFormula	varchar	period-names are binary in SQL; "translation":
		``SELECT
		    [MyDateFormula (Text)] =
		        REPLACE(
		          REPLACE(
		            REPLACE(
		              REPLACE(
		                REPLACE([MyDateFormula],
		                  CHAR(7), 'Y'), -- Year
		                  CHAR(6), 'Q'), -- Quarter
		                  CHAR(5), 'M'), -- Month
		                  CHAR(4), 'W'), -- Week
		                  CHAR(2), 'D'), -- Day
		                  CHAR(1), 'C') -- Current
		FROM [CRONUS$Record]```	

TableFilter	varbinary		
GUID	uniqueidentifier	{12345678-1234-1234-1234-1234567890AB}.	
Duration	bigint		
RecordID	varbinary		

STRYK System Improvement
Performance Optimization & Troubleshooting

With a "Linked Object" it is also mandatory to have **identical names** – table and fields - of the SQL site "View" and the NAV table.

Here it has to be regarded, if the NAV table property "**DataPerCompany**" is TRUE (default) NAV expects a SQL site table/view name starting with the COMPANYNAME; e.g. `CRONUS$MyViewAsLinkedObject`!

The NAV/SQL Performance Field Guide
Version 2009

Example (using the <u>Server</u> Date/Time in NAV):

SQL Server Site "View":

```
USE [Navision]
GO
CREATE VIEW [dbo].[Server DateTime] AS
  SELECT getutcdate() AS [Server DateTime]
GO
GRANT SELECT ON [dbo].[Server DateTime] TO public
```

NAV Table (Linked Object):

```
OBJECT Table 50000 Server DateTime
{
  OBJECT-PROPERTIES
  {
    Date=24.09.08;
    Time=12:00:00;
    Version List=SSI/Example;
  }
  PROPERTIES
  {
    DataPerCompany=No;
    LinkedObject=Yes;
    LinkedInTransaction=No;
  }
  FIELDS
  {
    { 1   ;   ;Server DateTime    ;DateTime         }
  }
  KEYS
  {
    {     ;Server DateTime                          }
  }
  CODE
  {

    BEGIN
    END.
  }
}
```

>> Using Temporary Tables

Most transactions in NAV are performed on "real", non **temporary tables**; a usual processing could look like this:

```
Record.SETRANGE(Field1, Value1);
IF Record.FINDSET(TRUE, FALSE) THEN
  REPEAT
    Record.Field02 := Value2;
    Record.MODIFY;
  UNTIL Record.NEXT = 0;
```

With this actually a cursor is created and each record is processed with an UPDATE command. These Updates result in changes of fields and records thus in pages, indexes, etc., which is work the SQL Server has to perform. Depending on the volume of the transaction, also physical I/O from and to the disks is necessary.
In case of an error all changes have to be undone – additional work for the server.

To speed up transactions it could be feasible to perform the transactions in temporary tables, proceeding like this:

1. Define Result-Set and **load** records into temporary table
2. Process temporary table to perform the **updates**
3. "**Dump**" the changes from the temporary table to the real one

This is recommended when <u>extensive</u> business logic has to be processed and many changes will occur (insertion, modification, deletion). The advantage is, that locks can only be established at the very end of the process – during the dump – so the probability of conflicts with other transactions is small.
Else locks would be established too early, maybe blocking other processes.

Further, as the result-set is processed on Client site, the SQL Server is somewhat relieved. This implicates that the client machine has to have sufficient memory and CPU power to process the transactions.

Load:

```
Record.SETRANGE(Field1, Value1);
IF Record.FINDSET(FALSE, FALSE) THEN
  REPEAT
    TmpRecord.INIT;
    TmpRecord := Record;
    TmpRecord.INSERT;
  UNTIL Record.NEXT = 0;
```

Dump:

```
IF TmpRecord.FINDSET THEN
  REPEAT
    Record.GET(TmpRecord.<Primary_Key>);
    Record := TmpRecord;
    Record.MODIFY;
  UNTIL TmpRecord.NEXT = 0;
```

Caution: If a record variable is declared as "temporary" this only affects this record. All other subsequently used records – e.g. with any Triggers – are not temporary by default!

With NAV 5.0 a new "Form" property was introduced: "**SourceTableTemporary**"
If this is set to TRUE, a temporary table could be used as SourceTable.

STRYK System Improvement
Performance Optimization & Troubleshooting

> Miscellaneous Issues

Other issues to mention briefly, for gaining better performance, are

- The used C/SIDE Version
- The DBCC PINTABLE feature
- NAV "Table Optimizer"

>> The used C/SIDE Version

C/SIDE (*Client/Server Integrated Development Environment*) – actually the NAV client – is establishing the connection to the SQL Server, it "translates" the **C/AL** (*C/SIDE Application Language*) code into SQL, sends the queries to the server and processes the result-set within the business logic.
Therefore, the used C/SIDE version has remarkable influence on the whole system: the better the C/AL code is translated, the faster could the SQL Server process the queries!
It is recommended to always use the **most recent C/SIDE version** to benefit from this improved translation and from additional methods/functions/etc.. Further, C/SIDE generates the SQL Server site SIFT triggers: the better the trigger-code, the better the SIFT performance!

It should mentioned, that the used C/SIDE version is actually independent from the version of the application objects; for example a 3.70 database application could run on a 4.00 C/SIDE version.

It is not necessary to run a "full" upgrade here, a **technical upgrade** (updating C/SIDE *program components*, not updating the *application objects*) is sufficient!

During such an upgrade NAV asks to "*convert*" the database. Well, the DB cannot be "converted" – it will always be a SQL DB – but this "*conversion*" actually re-writes the SIFT Triggers (thus rebuilding the SIFT structures completely) and changes some internal NAV tables.

See **Appendix B** for existing versions of MS Dynamics NAV and SQL Server.

>> The DBCC PINTABLE feature

Applies to SQL Server 2000 only

With **DBCC PINTABLE** a table could be marked to prevent SQL Server to flush it from the cache.
This could be feasible for small but "hot" tables (often used) to avoid disk I/O. Good candidates are "*No. Series*" or the "*Object*" table.

Example:

```
DECLARE @db_id int, @tbl_id int
SET @db_id = DB_ID('Navision')
SET @tbl_id = OBJECT_ID('Object')
DBCC PINTABLE (@db_id, @tbl_id)
```

Caution: If a table was pinned which occupies too much of available cache, the SQL Server cannot process anymore. Even though the table could be unpinned with **DBCC UNPINTABLE** it is necessary to **restart** the SQL Server to flush the cache!

Example:

```
DECLARE @db_id int, @tbl_id int
SET @db_id = DB_ID('Navision')
SET @tbl_id = OBJECT_ID('Object')
DBCC UNPINTABLE (@db_id, @tbl_id)
```

To get an overview about which tables are pinned, this query could be executed:

```
SELECT * FROM sysobjects
WHERE objectproperty(id, N'TableIsPinned') = 1
```

To determine the amount of RAM which is occupied by pinned tables, the command **DBCC MEMORYSTATUS** could be used; here in section 1 the "**INRAM**" entry shows the number of pages used for those tables. This figure multiplied with 8 gives the amount of cache used in Kilobytes.

Remark: DBCC PINTABLE and DBCC UNPINTABLE are not available anymore in **SQL Server 2005** as it was too problematic. The syntax is still valid but has no impact on the server.

>> NAV "Table Optimizer"

The standard NAV "**Table Optimizer**" is also doing some kind of maintenance as it recreates the indexes and cleans up the SIFT tables. As it is **resetting** all indexes to the sub-optimal NAV standard (the same happens when **modifying** a key/index in C/SIDE) it is not recommended to use it.

>> Index Hinting

With NAV it is possible to set hints about which index to use for specific queries (for details refer to the "*NAV Performance Troubleshooting Guide*").

1. Create configuration table:

```
USE [Navision]
GO

CREATE TABLE [$ndo$dbconfig] (config VARCHAR(1024))
GRANT SELECT ON [$ndo$dbconfig] TO [public]
```

2. Define Index Hint(s) for specific query:

```
INSERT INTO [$ndo$dbconfig] VALUES
('IndexHint=Yes;Company="CRONUS";Table="Item Ledger Entry";
Key="Item No.","Variant Code";Search Method="-+";Index=2')
```

With this hint, the index $2 will always be used for **FIND('-')** or **FIND('+')** methods on the **Item Ledger Entry**; company **CRONUS**.

Caution:
Normally SQL Server will pick the best index for a query, so overruling this by using Index Hinting could be problematic and should only be used when really necessary and all other kinds of optimizations failed!

Index Hinting is NOT recommended!

With NAV 4.00 Build **25143** "*Index Hinting*" is enabled by default! It should be deactivated by setting this flag in the ndodbconfig table:

```
INSERT INTO [$ndo$dbconfig] VALUES ('IndexHint=No')
```

>> Different Query Performance in SQL 2000 and 2005

In some cases it could happen, that queries which were performed good with SQL Server 2000 are executed worse in SQL Server 2005, requiring thousands of page Reads more.

If these bad Execution Plans were created due to filtering using the **LIKE** operator – which is caused by using wildcards in NAV – this problem cannot be solved with "*Index Hinting*" or using the *OPTION(RECOMPILE)* feature (e.g. via "*Plan Guides*").

The queries which are probably affected by this problem actually include a *LIKE* operator and a *comparison* operator as > (greater than) or < (less than) on the same field; e.g.

```
'SELECT   * FROM "Navision"."dbo"."CRONUS$Customer" WHERE
(("Search Name" LIKE @P1)) AND  "Search Name"<@P2 ORDER BY
"Search Name" DESC,"No_" DESC ',@p3 output,@p4 output,@p5 out-
put,N'@P1 varchar(30),@P2 varchar(30)','ME%','BEEF HOUSE'
```

Normally, with SQL Server 2005 the **comparison** operator would be used to perform the seek; instead of the **LIKE** operator (SQL Server 2000 behavior); thus more rows are returned, which is slower.

With SQL Server 2005 build **9.00.3200** Microsoft has introduced the new **trace-flag 4119**.

With 4119 enabled, SQL Server 2005 will use the **LIKE** operator to seek (SQL 2000 behavior).

See also http://support.microsoft.com/kb/942659/

Enable trace-flag:

```
DBCC TRACEON (4119, -1)
```

(or via Startup-Parameter **–T4119**)

Getting rid of Locks, Blocks and Deadlocks

> Locking Mechanisms

>> Lock Escalation

Both database servers – "native" and SQL Server - have in common, that a single record is locked **at the moment** when a write transaction occurs (FDB: **Version Principle**, FDB/SQL: **Optimistic Concurrency**).

If enhanced locks should be established on FDB the only available command is `LOCKTABLE` which locks the whole table.

As mentioned above, locking in SQL Server is much more complex, there are different types of locks as *Shared (S), Exclusive (X), Update (U) or Indented (I)* locks, and there are multiple levels of locking as *Row, Range, Page, Index, Key, Partition, Table*, etc..

SQL Server will always (mostly) start locking on the lowest level, the row – `RECORD-LEVELLOCKING` (ROW). But too many locks of a lower level have to be maintained, then the server Could *escalate* to a higher level. For example, thousands of ROW locks could be replaced by a single RANGE lock.
The idea of this escalation is to save system-resources and to reduce the administrative effort, so actually Lock Escalation is necessary to process large results fast.

The disadvantage is, that the higher the locking level, the higher is the risk of encountering blocking conflicts, which is just a matter of **probability**:
For example, if two process are ROW locking, the probability that both are locking the same record (e.g. "Sales Header" *record*) is much smaller; if they would lock the whole TAB (e.g. "Sales Header" *table*) the probability is much higher.

> **Reducing the locking-level reduces blocks**

"**Lock Escalation**" could be prevented by using the "**Always Rowlock**" database property (as mentioned above). Thus "*Always Rowlock*" is reducing the blocking probability, but at cost of system resources: with heavily used 32bit systems it could be problematic, with 64bit systems it should work fine (depends on Server Sizing).

>> Implicit Locking

As mentioned above, the principle of "**Optimistic Concurrency**" is also implemented in SQL Server. This means, that no locks are established, latest until a transaction (= write operation – INSERT/UPDATE/DELETE) happens.
For example, if a process is reading a "*Customer*" record only **Shared Locks** are set; other processes are still allowed to read this record, too.
If now a process starts to write to the record – e.g. changing the "*Address*" – the SQL Server will **implicitly** set an **Exclusive Lock**. Thus, if finally a second process starts to write to the record, too, an **error** message will be raised, telling something like "*Another user has changed the record …*".

Hence, all write-transactions will automatically result in **Exclusive Locks** on the affected resource, and also **Intended Exclusive Locks** on higher extents (see below).

>> Explicit Locking

"**Explicit Locking**" could be accomplished by using the `LOCKTABLE` command in NAV.

Have in mind that a `LOCKTABLE` does not instantly set a lock on a table:
it sets the **TRANSACTION ISOLATION LEVEL** to **SERIALIZABLE** (standard NAV isolation is **READUNCOMMITTED**) and adds the query hint `UPDLOCK` (U), transforming **Shared Locks** (S) into **Exclusive Locks** (X).

Hence, once a transaction is "serialized" the process can only read **committed data**, thus no "**Dirty Reads**" are possible; data *consistency* is granted.
The downside is, that the process gets blocked if trying to read **un-committed data** written by other processes.

`LOCKTABLE` prevents "*Dirty Reads*" but increases the risk of getting blocked. So when it is about using the `LOCKTABLE` command it is a matter of

Data Consistency versus Blocking Risk.

Have in mind that a **FINDSET(TRUE)** behaves like **LOCKTABLE**!

STRYK System Improvement
Performance Optimization & Troubleshooting

In some cases `LOCKTABLE` might be avoided; for example, in Codeunit 80 "*Sales Post*" there is this code block:

```
IF RECORDLEVELLOCKING THEN BEGIN
   IF WhseReceive THEN
      WhseRcptLine.LOCKTABLE;
   IF WhseShip THEN
      WhseShptLine.LOCKTABLE;
   DocDim.LOCKTABLE;
   IF InvtPickPutaway THEN
      WhseRqst.LOCKTABLE;
   SalesLine.LOCKTABLE;
   ItemChargeAssgntSales.LOCKTABLE;
   PurchOrderLine.LOCKTABLE;
   PurchOrderHeader.LOCKTABLE;
   GLEntry.LOCKTABLE;
   IF GLEntry.FINDLAST THEN;
END;
```

This might be sub-optimal. The idea of this code is to set a **semaphore** in place to avoid certain block-/deadlock situations. If purely relying on "**Optimistic Concurrency**" – means <u>assuming</u> the **probability** that e.g. the same "Sales Line" record is modified by two processes *at the same time* is very small – then the implicit ROW-lock level of SQL Server *could* be sufficient enough, so in a "perfect world" the remaining code could be like this:

```
IF RECORDLEVELLOCKING THEN BEGIN
     GLEntry.LOCKTABLE;
     IF GLEntry.FINDLAST THEN;
END;
```

If this is feasible or not depends on the *business processes* (remember: consistency vs. blocks)!

But still this causes blocks:
With this algorithm actually the last "*G/L Entry*" record is exclusively locked (ROW S ➔ U ➔ ROW X). If this happens, the whole data Page is flagged with an "***Intended Exclusive Lock***" (**PAG IX**). If this is the case, no other process is allowed to establish another ROW X lock on that Page (e.g. by inserting another record), hence, the second process is blocked.
Another serialized (LOCKTABLE) process is not even allowed to *read* this locked records – the process is blocked.

This and similar algorithms are common in all kinds of posting routines, e.g. whenever a new "Entry" is inserted. Hence, only one process could insert an Entry at one point in time – standard NAV has a natural limit of concurrent postings; there will always be blocking issues!

According to this it is important to **speed up** transactions (Code & Index Tuning) as much as possible. The faster the processing, the smaller is the probability to interfere with other transactions!

Another issue with this piece of code in CU 80: the "*G/L Entry*" is blocked, regardless if a "*G/L Entry*" record will be created or not (for example the posting of a "*Shipment*" does not necessarily create a "*G/L Entry*")!
So there could be a lot of blocking issues which could be avoided, e.g. by determining **IF** a "G/L Entry" will be created **BEFORE** locking:

```
IF RECORDLEVELLOCKING THEN BEGIN
  ...
  IF CreateGLEntry THEN BEGIN
    GLEntry.LOCKTABLE;
    IF GLEntry.FINDLAST THEN;
  END;
END;
```

The challenge here is to determine if the Entry will be created or not; this should be individually specified.

(BTW: Codeunit 90 is affected by the same thing! Hence, for example, posting a non-G/L Sales Shipment concurrent with a non-GL Purchase Shipment could encounter blocks!)

So the conclusion is actually this:

> Think carefully about which records to lock and to which extend!

STRYK System Improvement
Performance Optimization & Troubleshooting

> Using GUID

GUID = Globally Unique Identifier

The **GUID** is a 16 byte value which is generated by the OS and is **unique**. NAV supports the data-type GUID (see *C/SIDE Reference Guide*). In some cases GUID could be used to avoid blocking issues:

As mentioned above, many **Clustered Indexes** are based on an **increasing number** as *Entry No.* in various ledger entry tables. Here new data is always inserted at the "end" of the CI, on the last index page (leaf node). If this page is locked, no other process could write to this page, this process is blocked.

To avoid this, a new field of type GUID could be added to the table; the value could be generated in the *OnInsert* trigger using the **CREATEGUID** command. Now the *Clustered Index* could be created on this GUID field. This should be done with TSQL:

```
ALTER   TABLE [dbo].[CRONUS$Warehouse Entry]
DROP CONSTRAINT [CRONUS$Warehouse Entry$0]

ALTER   TABLE [dbo].[CRONUS$Warehouse Entry] WITH NOCHECK
ADD CONSTRAINT [CRONUS$Warehouse Entry$0]
PRIMARY KEY NONCLUSTERED ([Entry No_])

CREATE CLUSTERED INDEX [ssi_CIdx]
ON [dbo].[CRONUS$Warehouse Entry] ([UID])
WITH FILLFACTOR = 90
```

Hence, when inserting a record this page still will be blocked. But as the GUID numbers are not (!!!) in sequence, each insertion will take place on a **different** page, not only the last one! So the probability that multiple transactions try to write to the same index page is drastically minimized; blockings are reduced!

Important: While a CI/PK on "Entry No." (sequence) could have a Fill-Factor of 100% this is not recommended for GUID fields (non-sequence), here we need some free space to insert the index values. Hence, this GUID index is larger than the original one.

The NAV/SQL Performance Field Guide
Version 2009

Example – Physical Order "*Entry No. vs. GUID*":

Entry No.				GUID		
Entry No. (PK/CI)	**Value**		**Entry No. (PK/AI)**	**UID (CI)**	**Value**
1	A		5	{C49A537B-6EAF-419E-A119-0B39B2BF55FB}	E
2	B		9	{05D3594A-01BD-42ED-8484-35355EFCBA7B}	I
3	C		7	{BB2D7973-A8C1-414C-8786-3FB310B0C97F}	G
4	D		4	{AA9CAAEB-BF08-48D9-88E4-5C9BF1041271}	D
5	E		1	{E38C6700-A141-4F1B-8E0E-60AA8539E7D2}	A
6	F		2	{99CF77A6-DE3C-41A9-8CA4-65016886CD92}	B
7	G		8	{F95B7F86-6B4C-4562-91C4-C3F1F5977F12}	H
8	H		6	{9AD097ED-3C53-4F0C-903C-E03FD055EAE5}	F
9	I		3	{1F1A7CB0-2754-4E2A-A2D0-F14C50877777}	C

In both cases record number 8 is currently created. This causes an "exclusive lock" on the record (**ROW X**) and an "intended exclusive lock" on the page (**PAG IX**).

If the CI is based on "Entry No." it is not possible to write record number 9 until the previous transaction was committed – the process is blocked.

If the CI is based on the "GUID" field, record/page number 8 are also locked, but as record number 9 is inserted on a <u>different</u> physical page (!) this process does <u>not</u> get blocked!

> Using AutoIncrement

Mostly, any kind of *Entry Nos.* are created in NAV like this:

```
IF Record.FIND('+') THEN
  "Entry No." := Record."Entry No." + 1
ELSE
  "Entry No." := 1;
```

This is actually quite time consuming, as a query has to be sent to the SQL Server. The more records are in the table, the longer it could take.
By setting the field property "**AutoIncrement**" to Yes, the SQL Server would generate the number instead (**IDENTITY INSERT**), which is much faster!
As it is not possible to insert or modify AutoIncrement values, the field has to be initialized with zero, otherwise an error will be raised (depending on the user rights):

```
Record."Entry No." := 0;
```

The last value of this IDENTITY is actually **saved** on SQL Server site. Even if the table is emptied – all records deleted – the numbering will not restart with 1, it will continue from the last number. To reset IDENTITY it is necessary to drop/delete and re-create the table!

In some cases the "**GUID** CI Optimization" (see above) only is feasible together with **AutoIncrement**!

Caution: If using AI the SQL-site calculated number – for example the new "*Entry No.*" – is not instantly available in NAV, thus it might be necessary to **re-read** the new record!

> Forcing Row-Locking

As explained above, the setting "**Always Rowlock**" (NAV 4.00) does not really force row-locking – it adds the query hint ROWLOCK which prevents the escalation of locking levels.
Basically, too many row-locks require higher administrative costs than locking a higher instance as Page, so in most cases the "**lock-escalation**" is desired.

Nonetheless, in few cases it could be feasible to force row-locking. As the "*Always Rowlock*" feature is a global ON/OFF switch, it can't be used just to affect single tables or processes.

To force row-locking, it is necessary to just establish a lock on a single **record** - *which is not used anywhere else* - within a table, using an *HOLDLOCK* hint.
This would also establish **Intended Exclusive Locks** (IX) on the higher levels as Page (PAG IX) and Table (TAB IX).
As long as these intended locks are *hold*, no other process could establish an exclusive lock on these levels, e.g. a TAB X is not possible; in this case **no lock-escalation** could happen!

This locking of a "dummy" record should be performed via TSQL:

```
BEGIN TRANSACTION
SELECT * FROM dbo.[CRONUS$MyTable]
WHERE [No_] = FakeNo (HOLDLOCK)
```

This statement could be execute e.g. via a SQL Server Agent job when starting the service.

Even more convenient is the usage with **MS ADO** (see above) to temporarily establish this row-locking, because then the lock could be released if not needed anymore:

```
COMMIT TRANSACTION
```

It is not possible to provoke this row-locking within NAV via C/AL: the row-lock/holdlock has to be established by a **different process**, else the escalation could happen!

STRYK System Improvement
Performance Optimization & Troubleshooting

> Setting Lock Granularity

As mentioned, since NAV 4.00 it is recommended to use the "*Always Rowlock*" feature only with 64bit systems – with 32bit it might cause trouble and mostly should be disabled.

In previous versions (3.70 an earlier), C/SIDE will always send a ROWLOCK hint. To prevent this, one could implement a new table "**ndodbconfig**" to specify the lock granularity:

```
USE [Navision]
GO

CREATE TABLE [$ndo$dbconfig] (config VARCHAR(1024))
GRANT SELECT ON [$ndo$dbconfig] TO [public]
GO

INSERT INTO [$ndo$dbconfig] VALUES
('DefaultLockGranularity=Yes')
```

The NAV/SQL Performance Field Guide
Version 2009

> Block Detection

Resource Locks and blocks could be detected with certain **Performance Indicators** (e.g. *Lock Waits/sec* or *Average Waiting Time*), but these just tell *THAT* something was blocked, not *WHAT*.

This could be answered by **sp_who2**, **Activity Monitor**, etc. or via **TSQL** script:

```
USE [Navision]
GO

DECLARE @threshold int
SET @threshold = 1000   -- Milliseconds
WHILE 1 = 1 BEGIN   -- eternal loop
  IF EXISTS(SELECT TOP 1 NULL FROM master..sysprocesses
            WHERE [blocked] <> 0)
    SELECT
      [db] = db_name(s1.[database_id]),
      [waitresource] = ltrim(rtrim(s1.[wait_resource])),
      [table_name] = object_name(sl.rsc_objid),
      [index_name] = si.[name],
      s1.[wait_time],
      s1.[last_wait_type],
      s1.[session_id],
      session1.[login_name],
      session1.[host_name],
      session1.[program_name],
      [cmd] = isnull(st1.[text], ''),
      [query_plan] = isnull(qp1.[query_plan], ''),
      session1.[status],
      session1.[cpu_time],
      s1.[lock_timeout],
      [blocked by] = s1.[blocking_session_id],
      [login_name 2] = session2.[login_name],
      [hostname 2] = session2.[host_name],
      [program_name 2] = session2.[program_name],
      [cmd 2] = isnull(st2.[text], ''),
      [query_plan 2] = isnull(qp2.[query_plan], ''),
      session2.[status],
      session2.[cpu_time]
```

(continued next page)

```
-- Process Requests
  FROM sys.dm_exec_requests (NOLOCK) s1
  OUTER APPLY sys.dm_exec_sql_text(s1.sql_handle) st1
  OUTER APPLY sys.dm_exec_query_plan(s1.plan_handle) qp1
  LEFT OUTER JOIN sys.dm_exec_requests (NOLOCK) s2
    ON s2.[session_id] = s1.[blocking_session_id]
  OUTER APPLY sys.dm_exec_sql_text(s2.sql_handle) st2
  OUTER APPLY sys.dm_exec_query_plan(s2.plan_handle) qp2
  -- Sessions
  LEFT OUTER JOIN sys.dm_exec_sessions (NOLOCK) session1
    ON session1.[session_id] = s1.[session_id]
  LEFT OUTER JOIN sys.dm_exec_sessions (NOLOCK) session2
    ON session2.[session_id] = s1.[blocking_session_id]
  -- Lock-Info
  LEFT OUTER JOIN  master.dbo.syslockinfo (NOLOCK) sl
    ON s1.[session_id] = sl.req_spid
  -- Indexes
  LEFT OUTER JOIN sys.indexes (NOLOCK) si
    ON sl.rsc_objid = si.[object_id] AND sl.rsc_indid = si.[index_id]
  WHERE s1.[blocking_session_id] <> 0
      AND (sl.rsc_type in (2,3,4,5,6,7,8,9)) AND sl.req_status = 3
      AND s1.[wait_time] >= @threshold
  WAITFOR DELAY '000:00:05' -- duration between checks
END
```

As result it is shown, which processes are involved – **who is blocked by whom** – and which was **the locked resource** causing the conflict!

This could be used "online" via Query Analyzer / Management Studio; or as SQL Server Agent job: here it is feasible to trigger the job <u>automatically</u> via an *Alert* (monitoring "**SQL Server: General Statistics: Processes blocked**") and write the output into a file or table. In this case the <u>loop</u> has to be removed!

> Deadlocks

A **Deadlock** occurs if two or more processes try to access blocked resources which leads to each process is waiting for the other to release the lock:

Process 1	Process 2	Status
Read Resource #1	Read Resource #2	
Lock Resource #1	Lock Resource #2	
Read Resource #2		Process 1 is blocked, waiting for Resource #2
	Read Resource #1	Process 2 is blocked, waiting for Resource #1
		Each Process is waiting for the other: **Deadlock**

While with the "native" database server (FDB) this situation was shown as an "*application hang*" which forced to kill the client with *CTRL+ALT+DEL*, the SQL Server is able to resolve this by terminating one of the processes. So, the "*winning*" process could continue, the "*victim's*" transaction is rolled back:

Process 1	Process 2	Status
...	...	Each Process is waiting for the other: **Deadlock**
Victim	Winner	Process 1 is rolled back; Lock on Resource #1 is released
	Read Resource #1	

When deciding which process is "*Winner*" or "*Victim*" the SQL Server regards the CPU Time spent for the processes and the number of Pages which were changed and the **DEADLOCK_PRIORITY** (if defined). Finally the server kills the process which has done minor changes, thus limiting the "*damage*".

Deadlocks can be avoided by

- Avoiding Blocks
- Index Optimization
- Serialization
- Common Locking Order
- Transaction Speed

>> Index Optimization

The index optimization – as previously described – has also major impact on Deadlocks. Many DL do not occur due to locking conflicts on record- or table-level, they could be caused by conflicts on **index-level**. If "bad" indexes are used, resulting in **Index Scans** instead of *Index Seeks*, the probability of encountering a DL is increased.
An optimized index structure is actually a prerequisite for avoiding deadlocks.

>> Serialization / Semaphore

This means to block a common master resource – a record which is Read by other processes - when starting the transaction, like

```
MasterRecord.LOCKTABLE;
MasterRecord.GET;
MasterRecord.MODIFY;
```

Well, this prevents getting Deadlocks, but actually <u>decreases</u> performance dramatically
 High performance usually also means to have a high level of **parallelism**, <u>not</u> serialization.

Not recommended!

The NAV/SQL Performance Field Guide
Version 2009

>> Common Locking Order

If all transactions establish locks on resources in the same order no conflict could occur. The "*Performance Troubleshooting Tools*" provide features to define **Locking Order Rules** and to investigate processes for **Violation** and **Potential Deadlocks**.

The locking order is defined either by **explicit locking** mechanisms, e.g. LOCKTABLE, or **implicit locking**, e.g. INSERT, MODIFY or DELETE.
A specific locking order could be forced by using LOCKTABLE commands in the same sequence within potentially conflicting processes; e.g.

Process A	Process B
Record1.LOCKTABLE;	Record1.LOCKTABLE;
Record2.LOCKTABLE;	Record2.LOCKTABLE;

This would actually **serialize** the transactions, the processes will encounter locks but without being deadlocked. Not really recommended …

>> Transaction Speed

Most optimizations are aiming for **increasing speed** of transactions. The faster a transaction runs – the **shorter the duration** of locks – the lower is the **probability** there will be conflicts with other transactions.
This could be gained by optimized C/AL code and improved indexes!

>> Investigating Deadlocks

With the "*Client Monitor*" it is possible to analyze locking orders and - according to this - potential deadlocks. To "encounter" a Deadlock two <u>or more</u> processes could be involved, but all the various circumstances impossibly could be predicted.
So, it is necessary to analyze Deadlocks (DL) when they **occur**, to fix the reason for it and avoid it in future.

Deadlocks could be analyzed with

- SQL Server Error Log
- SQL Server Alert / Windows Performance Monitor
- SQL Profiler

SQL Server Error Log

To get detailed information about DL in the **Error Log** it is necessary to activate certain **Trace Flags**:

SQL Server 2000:

Flag	Description
1204	Returns the type of lock participating in the deadlock and the current command affect by the deadlock.
[1205]	Returns more detailed information about the command being executed at the time of a deadlock. Caution: in some cases too much data is written into the Error Log.
3605	Sends trace output to the error log.

SQL Server 2005/2008:

Flag	Description
1204	Returns the type of lock participating in the deadlock and the current command affect by the deadlock.
1222	Returns the resources and types of locks that are participating in a deadlock and also the current command affected, in an XML format that does not comply with any XSD schema.

The Trace Flags could be activated either by adding them as **Startup Parameter** (-T<Number>) or via TSQL:

```
DBCC TRACEON (<Number>, -1)
```

Deactivation with

```
DBCC TRACEOFF (<Number>, -1)
```

The status about enabled Trace Flags could be displayed with

```
DBCC TRACESTATUS (-1)
```

The Error Log could be viewed via Management Studio/Enterprise Manager or via TSQL:

```
EXEC master..xp_readerrorlog
```

The NAV/SQL Performance Field Guide
Version 2009

SQL Server Alert / Windows Performance Monitor

With the **Performance Monitor** a precise Deadlock analysis could not be performed. Anyway, the indicator "**Deadlocks/sec**" indicates that there are deadlocking problems and could be used for a further exploration of the SQL Error Log.

SQL Profiler

Using the **SQL Profiler** to trace the occurrence of Deadlocks is the most efficient method in addition to the analysis of the Error Log. Especially SQL Server 2005 provides improved features!

Recommended "Events" for tracing Deadlocks:

SQL Server 2000	SQL Server 2005/2008
Lock: Deadlock Lock: Deadlock Chain	*Lock: Deadlock Graph*

With SQL 2005+ the "**Deadlock Graphs**" can be exported into a **XDL** file. This is actually a **XML** file which could be processed and analyzed with any XML-capable-application; e.g. MS Excel!

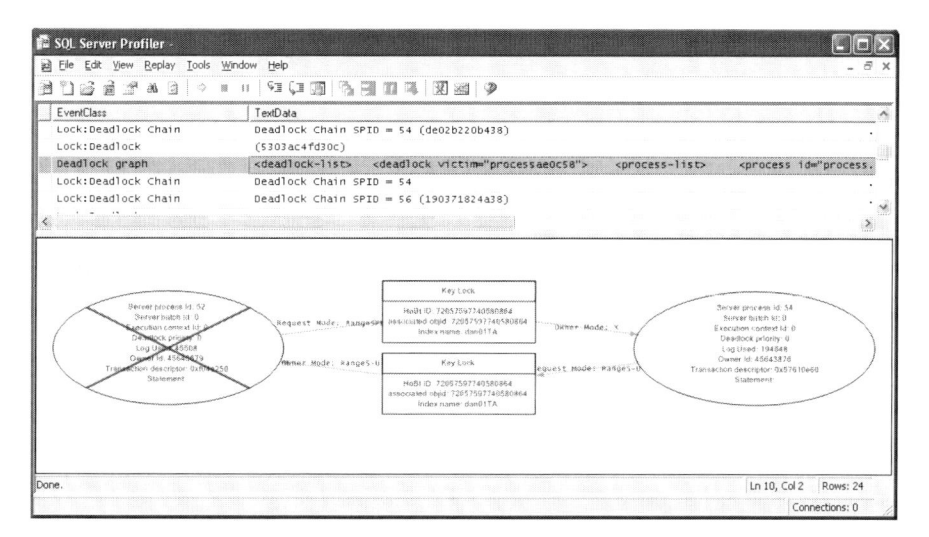

Page 133

STRYK System Improvement
Performance Optimization & Troubleshooting

As it is not feasible to run a permanent trace while using the Profiler online, it is recommended to export the Trace setup into a **TSQL script**.
This script could be started automatically as SQL Server Agent job (start with service) and **writing the results into a file** for further analyze.

```
DECLARE @rc INT
DECLARE @TraceID INT
DECLARE @maxfilesize BIGINT
SET @maxfilesize = 500
EXEC @rc = sp_trace_create @TraceID output
          ,0, N'c:\ssi_Deadlock_Trace', @maxfilesize, NULL
IF (@rc != 0) GOTO error
DECLARE @on BIT
SET @on = 1
EXEC sp_trace_setevent @TraceID, 148, 11, @on
EXEC sp_trace_setevent @TraceID, 148, 12, @on
EXEC sp_trace_setevent @TraceID, 148, 14, @on
EXEC sp_trace_setevent @TraceID, 148, 1, @on
DECLARE @intfilter INT
DECLARE @bigintfilter BIGINT
EXEC sp_trace_setstatus @TraceID, 1
SELECT TraceID=@TraceID
GOTO finish

error:
SELECT ErrorCode=@rc

finish:
GO
```

Once a trace was started via job, it could be monitored, stopped and removed via TSQL:

```
-- Status Information
SELECT * FROM fn_trace_getinfo(0)

DECLARE @TraceID int
SET @TraceID = <ID>

-- End Trace
EXEC sp_trace_setstatus @TraceID, 0

-- Delete Trace
EXEC sp_trace_setstatus @TraceID, 2
```

Database Maintenance

As mentioned before, it is crucial to run periodic maintenance on SQL Server databases. To setup this maintenance a SQL Server "**Maintenance Plan**" could be used, or SQL Server Agent **jobs** which are executing specific procedures etc..

> Maintenance Plan

Maintenance Plans (MP) could be used to setup, configure & schedule different kinds of optimization, as statistic updates, consistency checks or backups (for details refer to "*Books Online*" of SQL Server).

SQL Server 2000:

The disadvantage of MP in SQL Server 2000 is, that it cannot be configured to process on single tables or indexes, so for specific/individual maintenance it is not sufficient.
The MP actually creates several jobs which execute the MP via the `sqlmaint.exe` – indirectly using SQL procedures! A known problem of this `sqlmaint.exe` is, that it could crash once in a while.

Hence, the MP feature should be used to establish some basic periodic optimization. More specific maintenance should be performed with SQL Server Agent Jobs.

STRYK System Improvement
Performance Optimization & Troubleshooting

SQL Server 2005/2008:

Since SQL Server 2005 the MP feature has been remarkably improved; it is strongly **recommended** to use (if no better tools are available ☺) it and combine it with individual tasks!

Example (Backup):

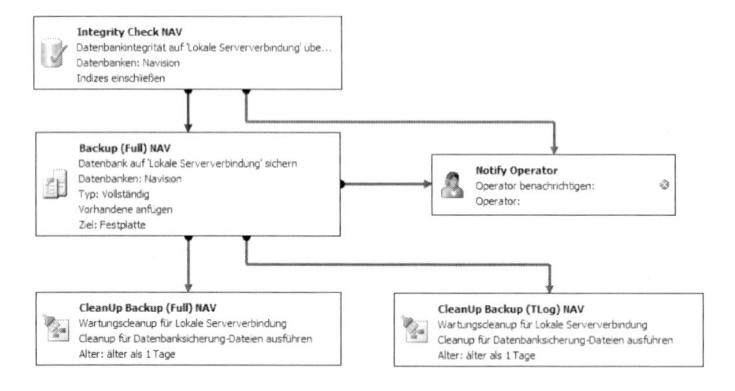

Example (Re-Index):

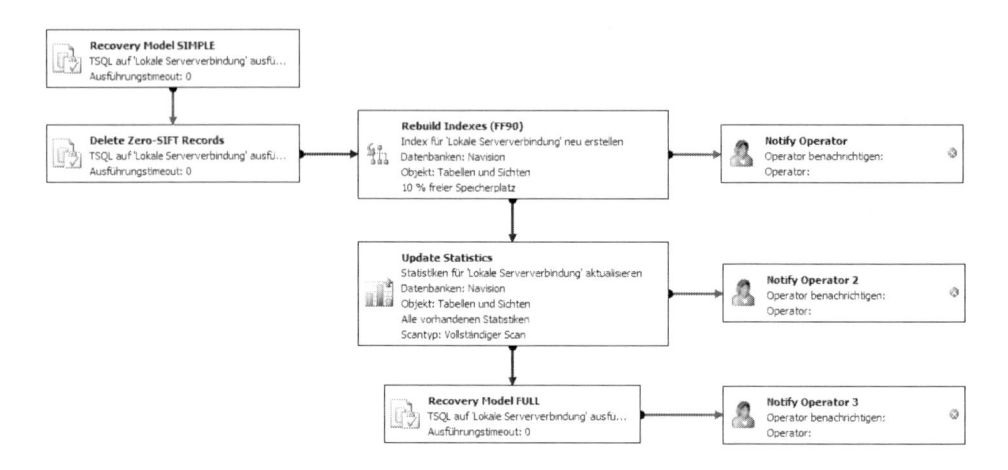

> SQL Server Agent Jobs

In the chapter about "*Database Configuration*" it was recommended to disable all kinds of "*Auto.*" features. To compensate this, it is necessary to execute several SQL procedures etc. (instead or in addition to *Maintenance Plan*). The following gives few examples:

Statistic Maintenance:

```
EXEC sp_updatestats
GO
EXEC sp_createstats 'indexonly'
```

Integrity Check:

```
DBCC CHECKDB ('<database>')
```

With repair (requires *Single User* mode):

```
DBCC CHECKDB ('<database>', REPAIR_FAST)
```

or

```
DBCC CHECKDB ('<database>'), REPAIR_REBUILD)
```

Cycle Error Log:

For troubleshooting purposes it is more convenient to have one **Error Log** per Day. When "*cycling*" the log (at 00:00 hours) a new log is created; according to this the number of logs to keep has to be configured, e.g. 30 (to keep one log per day for one month).

```
EXEC sp_cycle_errorlog
```

STRYK System Improvement
Performance Optimization & Troubleshooting

Index Defragmentation:

(SQL 2005/2008; see http://msdn.microsoft.com/en-us/library/ms188917.aspx)

```
USE [Navision]
GO
SET NOCOUNT ON;
DECLARE @objectid int;
DECLARE @indexid int;
DECLARE @partitioncount bigint;
DECLARE @schemaname nvarchar(130);
DECLARE @objectname nvarchar(130);
DECLARE @indexname nvarchar(130);
DECLARE @partitionnum bigint;
DECLARE @partitions bigint;
DECLARE @frag float;
DECLARE @command nvarchar(4000);
SELECT object_id AS objectid, index_id AS indexid,
       partition_number AS partitionnum,
       avg_fragmentation_in_percent AS frag
INTO #work_to_do
FROM sys.dm_db_index_physical_stats (
    DB_ID(), NULL, NULL , NULL, 'LIMITED')
WHERE avg_fragmentation_in_percent > 10.0 AND index_id > 0;
DECLARE partitions CURSOR FOR SELECT * FROM #work_to_do;
OPEN partitions;
WHILE (1=1)
    BEGIN;
        FETCH NEXT
           FROM partitions
           INTO @objectid, @indexid, @partitionnum, @frag;
        IF @@FETCH_STATUS < 0 BREAK;
        SELECT @objectname = QUOTENAME(o.name),
               @schemaname = QUOTENAME(s.name)
        FROM sys.objects AS o
        JOIN sys.schemas as s ON s.schema_id = o.schema_id
        WHERE o.object_id = @objectid;
        SELECT @indexname = QUOTENAME(name)
        FROM sys.indexes
        WHERE  object_id = @objectid AND index_id = @indexid;
        SELECT @partitioncount = count (*)
        FROM sys.partitions
        WHERE object_id = @objectid AND index_id = @indexid;
```

(continued next page)

```
        IF @frag < 30.0
            SET @command =
              N'ALTER INDEX ' + @indexname + N' ON ' +
              @schemaname + N'.' + @objectname + N' REORGANIZE';
        IF @frag >= 30.0
            SET @command =
              N'ALTER INDEX ' + @indexname + N' ON ' +
              @schemaname + N'.' + @objectname + N' REBUILD';
        IF @partitioncount > 1
            SET @command =
              @command + N' PARTITION=' +
              CAST(@partitionnum AS nvarchar(10));
        EXEC (@command);
        PRINT N'Executed: ' + @command;
    END;
CLOSE partitions;
DEALLOCATE partitions;
DROP TABLE #work_to_do;
GO
```

(This script maintains indexes on basis of the fragmentation scale as described above)

Clean Up Backup History:

Delete old entries from the Backup History:

```
DECLARE @date datetime
SET @date = getdate() - 30    -- delete older than 30 days
EXEC sp_delete_backuphistory @date
```

> Backup Strategy

According to Maintenance it is also necessary to run a sufficient **backup strategy** and/or **failover solutions**. Scheduling backups optimally avoids performance-drops due to conflicts with other transactions and it reduces the size of the Transaction Log, which also speeds up the system.

When establishing a backup strategy it is most important to define the demands regarding "**acceptable server down-times**" and "**acceptable loss of data**". For details refer to the "*Installation & System Management*" (w1w1isql.pdf) and "*Making Backups*" (w1w1bkup.pdf) documents for MS Dynamics NAV.

With SQL Server it is not recommendable to use the NAV built-in backup feature! This could not be performed automatically, it takes longer (especially if the database is large) and the restore could be problematic.
It could be feasible to perform backups using the "native" feature to save the NAV **objects** (if objects are saved in a FBK file they could be re-imported/restored from that!), e.g. also automatically using third party utilities.

Depending on the demands and capabilities, a reasonable backup strategy could look like this:

- **Full Backup** every 24 hours, e.g. after a business day, 18:00
- **Differential Backup** every 2 to 6 hours
- **Transaction Log Backup** every 15 to 60 minutes

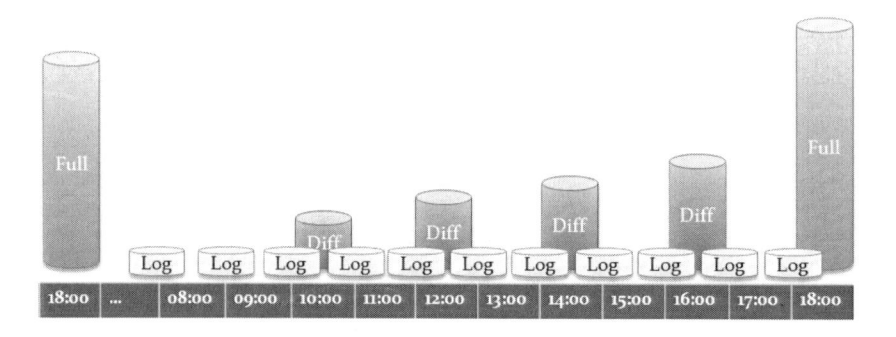

Caution: In case of disaster-recovery it is necessary to backup the last used Transaction Log to restore the database until the last committed transaction. If this is not available, the restore "ends" with the previous TLog backup.

Hence, basically the frequency of TLog backups "defines" how much data could be lost.

To avoid conflicts of backup- and maintenance jobs it is recommended to document the schedule, e.g. within an Excel sheet; for example:

Time	Mo	Tu	We	Th	Fr	Sa	Su
00:00	Cycle Log	Cycle Log	Cycle Log	Cycle Log	Cycle Log	Cycle Log	Cycle Log
01:00							
02:00							
03:00							
04:00							
05:00	Stats	Stats	Stats	Stats	Stats		Stats Full
06:00							
...							
11:00						SIFT	
12:00	Bak Diff	Bak Diff	Bak Diff	Bak Diff	Bak Diff	Growth	
13:00						ReIndex	
...							
18:00							
19:00	Bak Full	Bak Full	Bak Full	Bak Full	Bak Full	Bak Full	Bak Full
20:00							
21:00							
22:00							
23:00							

Backup with SAN or other external tools

Some SAN provide so called "**snapshot**" solutions to make very quick backups of complete LUN (see above for more about SAN). Have in mind that SQL Server writes "dirty pages" to disk, means data-changes from non-committed transactions are physically stored. To maintain the **database integrity** it is crucial that these "snapshot" tools – or other external backup tools – are aware of this and can deal with it.
If you want to use such tools make sure that the manufacturer grants compatibility with SQL Server!

STRYK System Improvement
Performance Optimization & Troubleshooting

Parameter Sniffing

When C/SIDE is sending a query to the SQL Server it is doing this with a **Parameterized Remote Procedure Calls** – means the SQL statement includes placeholders for some parameters, example (excerpt from Profiler trace):

```
RPC:Completed      declare @P1 int
set @P1=180150216
declare @P2 int
set @P2=2
declare @P3 int
set @P3=2
declare @P4 int
set @P4=1
exec sp_cursoropen @P1 output, N'SELECT * FROM "navi-
sion"."dbo"."CRONUS$Document Dimension" WITH (UPDLOCK, ROWLOCK)    WHERE (("Ta-
ble ID"=@P1 OR "Table ID"=@P2)) AND (("Document Type"=@P3)) AND (("Document
No_"=@P4)) ORDER BY "Table ID","Document Type","Document No_","Line
No_","Dimension Code" OPTION (FAST 25)', @P2 output, @P3 output, @P4 output,
N'@P1 int,@P2 int,@P3 int,@P4 varchar(20)',  36,  37,  1,  '5676158191'
select @P1, @P2, @P3, @P4   319       2          234        32503      7789
```

When this query is executed the first time, SQL Server stores the used execution plan in its procedure cache. If now a similar query is sent, with just different parameters, SQL Server does not necessarily create a new execution plan, it will consider the first parameters as representative for the following queries.
This feature (it's not a bug ☺) is called "**parameter sniffing**".
Hence, if the first query caused a "bad" execution plan, e.g. *Clustered Index Scans*, all following queries are causing the same trouble, even though a better plan could be generated.
(This could happen due to insufficient "*cursor handling*" – see above "**Dynamic Cursor**", "**Fast Forward Cursor**" and "**SQL Index**" property)

The tricky thing is to detect this "parameter sniffing": when copying the query to the Query Analyzer / Management Studio one usually replaces the parameters to create a "clean" query, like

```
SELECT   * FROM "navision"."dbo"."CRONUS$Document Dimension" WITH (UPDLOCK, ROW-
LOCK)
WHERE (("Table ID"=36 OR "Table ID"=37)) AND (("Document Type"=1)) AND (("Docu-
ment No_"='5676158191'))
ORDER BY "Table ID","Document Type","Document No_","Line No_","Dimension Code"
OPTION (FAST 25)
```

This **ad-hoc query** would result in a different execution plan, thus the original sub-optimal performance could not be reproduced!

Another way to find potential "*Parameter Sniffing*" issues is using this Script (from MS, applies to SQL Server 2005/2008 only):

```
SELECT
  st.text,
  SUBSTRING(st.text, (qs.statement_start_offset / 2) + 1,
  ((CASE statement_end_offset
      WHEN -1 THEN DATALENGTH(st.text)
      ELSE qs.statement_end_offset
    END
  - qs.statement_start_offset) / 2) + 1) AS statement_text,
  execution_count,
  CASE
    WHEN execution_count = 0 THEN NULL
    ELSE total_logical_reads / execution_count
  END AS avg_logical_reads,
  last_logical_reads,
  min_logical_reads,
  max_logical_reads,
  max_elapsed_time,
  CASE
    WHEN min_logical_reads = 0 THEN null
    ELSE max_logical_reads / min_logical_reads
  END AS diff_quota
FROM sys.dm_exec_query_stats AS qs
CROSS APPLY sys.dm_exec_sql_text(qs.sql_handle) AS st
ORDER BY max_logical_reads DESC
```

STRYK System Improvement
Performance Optimization & Troubleshooting

Compare the **min_logical_reads** with the **max_logical_reads**. If a query plan generates a few reads sometimes, and then it generates many reads other times the query is probably a candidate for RECOMPILE (SQL Server 2005).

To force the recompilation of an Execution Plan from NAV site the table **ndodbconfig** could be used (see above):

1. Create configuration table:

```
USE [Navision]
GO

CREATE TABLE [$ndo$dbconfig] (config VARCHAR(1024))
GRANT SELECT ON [$ndo$dbconfig] TO [public]
GO
```

2. Define RECOMPILE(s) for specific tables:

```
INSERT INTO [$ndo$dbconfig]
VALUES
('UseRecompileForTable="G/L Entry"; Company="CRONUS";
RecompileMode=2;')
```

Recompile Modes:

Mode	Description
0	Don't use OPTION (RECOMPILE)
1	Use OPTION (RECOMPILE) when browsing the table in a form (default)
2	Use OPTION (RECOMPILE) operations caused by the C/AL code
3	Always use the OPTION (RECOMPILE)

With SQL Server 2005/2008 this could also be solved by generating "**Plan Guides**" (aka "**Plan Freezing**") to force the re-compilation of Execution Plans:

```
EXEC sp_create_plan_guide
  @name = N'SSI_Guide1',
  @stmt = N'SELECT * FROM "navision"."dbo"."CRONUS$Document Dimen-
sion" WITH (UPDLOCK, ROWLOCK)  WHERE (("Table ID"=@P1 OR "Table
ID"=@P2)) AND (("Document Type"=@P3)) AND (("Document No_"=@P4)) ORDER
BY "Table ID","Document Type","Document No_","Line No_","Dimension
Code" OPTION (FAST 25)',
  @type = N'SQL',
  @module_or_batch = NULL,
  @params = N'@P1 int,@P2 int,@P3 int,@P4 varchar(20)',
  @hints = N'OPTION (RECOMPILE)'
```

View Plan Guides:

```
SELECT * FROM sys.plan_guides
```

Managing Plan Guides:

```
EXEC sp_control_plan_guide @operation = N'ENABLE ALL'
EXEC sp_control_plan_guide @operation = N'DISABLE ALL'
EXEC sp_control_plan_guide @operation = N'DROP ALL'
EXEC sp_control_plan_guide @operation = N'ENABLE',@name=N'SSI_Guide1'
EXEC sp_control_plan_guide @operation = N'DISABLE',@name=N'SSI_Guide1'
EXEC sp_control_plan_guide @operation = N'DROP',@name=N'SSI_Guide1'
```

Using RECOMPILE is only possible with SQL Server 2005, build 9.00.**3152** or higher recommended (see **Appendix B**).

If recompiling is not sufficient – or possible (SQL Server 2000) -, then the re-creation of an execution plan could be forced by emptying the procedure cache using

DBCC FREEPROCCACHE

STRYK System Improvement
Performance Optimization & Troubleshooting

High Availability & Failover Strategies

The following should give just a brief overview about available options. It has to be thoroughly discussed which on is feasible for each individual NAV/SQL scenario.

More information: http://msdn.microsoft.com/en-us/library/ms190202.aspx

> Failover Clustering

Applies to SQL Server 2000, 2005 and 2008

A **Failover Cluster** requires two or more nodes. In an **Active/Passive Cluster**, there is an **active node** – e.g. SQL Server – which is performing the actual processing; the other node – the **passive** – one, is running in stand-by mode ("*Hot Standby*").
A third node, so called **witness** is monitoring both nodes.
The database is stored on <u>one</u> **storage-/disk-subsystem** which is actually attached to the active <u>and</u> passive node.
If the active node fails, the witness will recognize this instantly and initiate a **failover-switch** to the formerly passive node, which then becomes the active node.
This is actually the most common Failover Cluster scenario. There is also the option of an **Active/Active Cluster**: here multiple nodes are performing the actual processing; hence a **load balancing** is accomplished.

Advantages	Disadvantages
Minimal Latency	Expensive (Software requirements)
Automatic failover	No redundant storage
Load-Balancing possible (A/A)	
Data Integrity	

> Database Mirroring

Applies to SQL Server 2005 and 2008

Actually the **Database Mirroring** is the "successor" of the *Transaction Log Shipping*. Here, the main SQL Server – the **principal** – transmits the **Transaction Log information** to a second SQL Server – the **mirror** – which is reprocessing the information. These processing-instructions are transferred from "*RAM to RAM*", so to speak, thus no physical files are copied or moved.
The mirrored database (on a separate disk-subsystem) is set into a permanent restore mode (**NORECOVERY**) ("*Warm Standby*"), constantly restoring the TLog information, hence this database is not available.

Mirroring could be performed in two variants:
- **Synchronous Mirroring** (aka "*High Security Mirroring*")
- **Asynchronous Mirroring** (aka "*High Performance Mirroring*") (Enterprise Edition only)

Synchronous Mirroring:

The principal sends the transaction data to the mirror and waits for the **commit**. Only if the commit is received from the mirror, then the principals commits the transaction, too. Hence, the client transactions has to wait for two commits: first from the mirror, then from the principal.
Thus, when processing a huge transaction volume too much time might be required to receive the commits and performance would be degraded.
The whole mirroring could be monitored by a "**witness**" – a third SQL Server (this could be an "*Express Edition*") – which could perform an automatic failover if the principal fails.

Advantages	Disadvantages
Minimal Latency	Cheap – Standard SQL feature
Automatic failover (Witness)	Degraded Performance
Data Integrity	Mirror DB not accessible
Redundant Storage	

STRYK System Improvement
Performance Optimization & Troubleshooting

Asynchronous Mirroring:

The principal **commits** the transactions <u>instantly</u> to the client and sends the TLog data asynchronously to the mirror. Hence, the principal's performance is <u>not</u> affected by waiting for the mirror.

If the mirroring is interrupted, the **data integrity** of the mirror could be affected – the principal's and the mirror's databases are out of sync.

A witness cannot be involved, thus an automatic failover is not possible.

Advantages	Disadvantages
Small Latency	Expensive – Enterprise SQL feature
High Performance	Manual/Semi-Automatic Failover
Redundant Storage	Data Loss is possible
	Mirror DB not accessible

> Transaction Log Shipping

Applies to SQL Server 2000, 2005 and 2008

Transaction Log Shipping is actually the predecessor of *Database Mirroring*. Here the Transaction Log data is **backed up** into a TRN file. Then this TRN file is **copied** or moved to the remote SQL Server which is restoring it, thus reprocessing the data. A "**witness**" could be involved to monitor the LS.

Hence, with Log Shipping the TLog data is transferred "*File to File*" while Mirroring transmits "*RAM to RAM*".

The remote server's database is also in **STANDBY** mode, thus the database could be used in "*read only*" mode ("*Warm Standby*").

If the TRN transfer is interrupted the remote database gets out of sync.

Actually Log Shipping is quite similar to "*Asynchronous Database Mirroring*", so it could be considered a cheaper alternative. The difference is, that with Mirroring each transactions is transferred <u>instantly</u>; with Log shipping the transfer speed depends on the frequency of TLog backups and the copy process.

Advantages	Disadvantages
Medium Latency	Cheap – Standard SQL feature
Good Performance	Manual/Semi-Automatic Failover
Redundant Storage	Data Loss is possible
Remote DB accessible (read only)	

> Database Snapshots

Applies to SQL Server 2005 and 2008

Actually **Database Snapshots** should <u>not</u> be considered a failover solution. But this an ideal feature to complement the failover solutions **Database Mirroring** or **Transaction Log Shipping**.
With those failover solutions the remote/mirror databases are in permanent restore mode, thus not accessible.
But it is possible to generate a *Database Snapshot* from such a database – and this snapshot is <u>accessible</u> (read only), so it could be used e.g. for reporting purposes!
A snapshot – simplified - just saves some **meta-data** of the database, so it does not physically contain any data etc.. Even though a snapshot allocates disk-space, these files are actually almost empty in the beginning. Only if data is changed in the "*source database*" of the snapshot then the **previous Page** (the data before the change) will be physically saved. Hence the more data is changed, the more data has to be saved within the snapshot.
(That's why it is recommended to create snapshots just for temporary purposes and to delete them afterwards)

Example:

```
USE [master]
GO
CREATE DATABASE Navision_Snapshot ON
( NAME = Navision_Data,
  FILENAME = 'D:\Snapshots\Navision_Snapshot_Data.ss'),
( NAME = Navision_1_Data,
  FILENAME = 'D:\Snapshots\Navision_Snapshot_1_Data.ss')
AS SNAPSHOT OF Navision
GO
```

> Replication

Applies to SQL Server 2005 and 2008

Well, basically the **Replications** primary purpose might not be failover/high availability reasons – with Replication data excerpts should be transferred **to** remote databases but could also **receive/merge** data from remote servers.

There are different types of replication, which will not be discussed here (see "*Books Online*" about details):
- **Snapshot Replication**
- **Transactional Replication**
- **Merge Replication**

The **Transactional-Replication** could be feasible to establish remote failover databases ("*Warm Standby*").
The very difference to *Log Shipping* or *Mirroring* is this:
The remote database is accessible (read and write). But the data is not transferred as TLog data – the data is not *re-processed* on the remote server – the data is directly inserted, updated or deleted!
Hence, the transfer volume and the load on the remote server could be remarkably higher!

Advantages	Disadvantages
Medium Latency	Cheap – Standard SQL feature
Medium Performance	Manual/Semi-Automatic Failover
Redundant Storage	High Transfer Volume (I/O)
Remote DB is accessible	Schema Changes problematic (NAV Object changes cannot be replicated)

The NAV/SQL Performance Field Guide
Version 2009

The following should give an example how multiple failover/availability solutions could be combined:

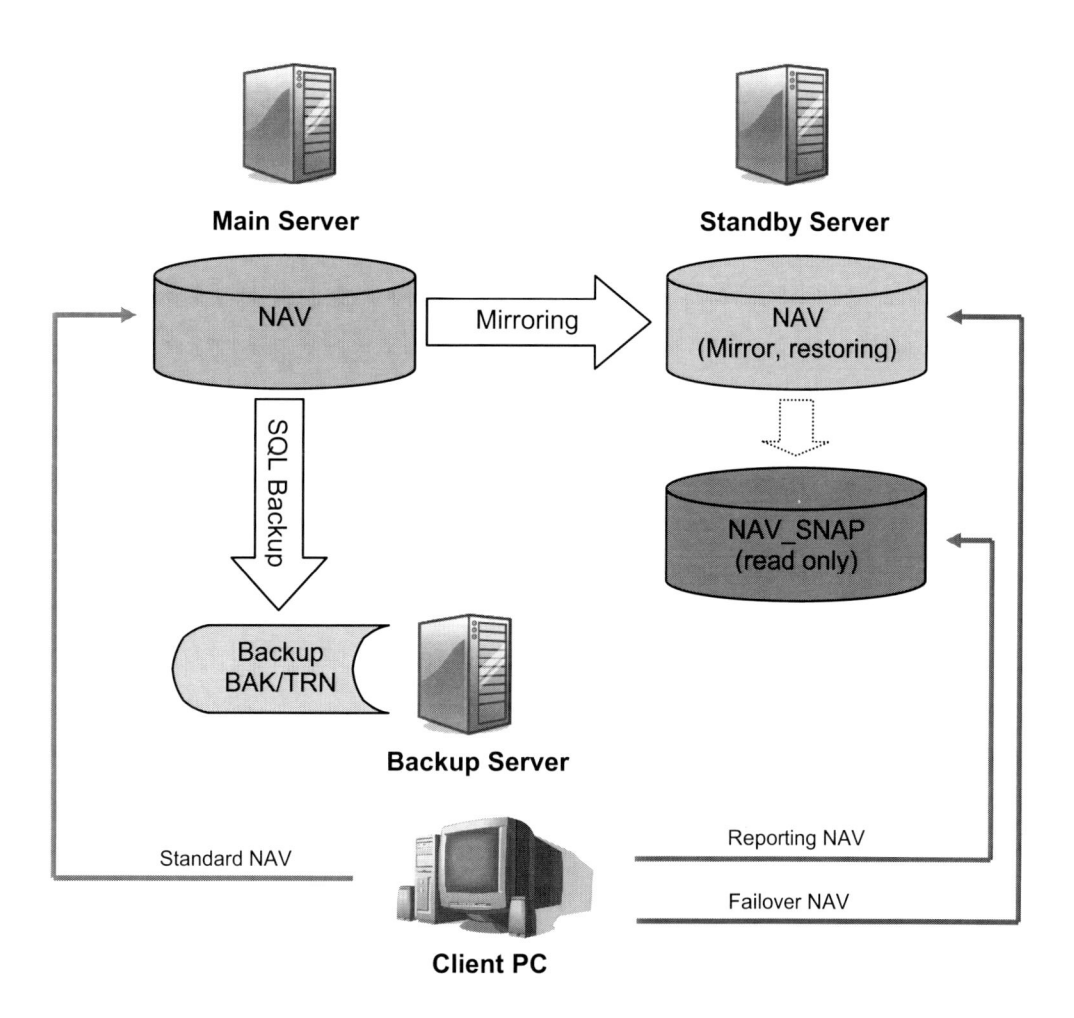

Miscellaneous

> Ghost Cleanup

When monitoring the "Transactions/sec" within the Performance Monitor it could occur, that even on a completely idle system without any user connections or transactions, a periodic peak is shown (looks like a "heart beat").
This is caused by the so called **GHOST CLEANUP** process:
When deleting data from the database, SQL Server marks those objects (rows, columns) to be physically deleted afterwards. This asynchronous deletion is performed by the system process GHOST CLEANUP.
Nothing to worry about, this actually improves the performance.

> Named Pipes

To connect a NAV client with SQL Server usually the **TCP**/IP protocol is used; which is actually quite feasible when working in a network.
When connecting to the SQL Server with a NAV client, Management Studio, etc. directly on the server machine itself, performance could be improved by using the **Named Pipes** protocol: here there are no network roundtrips required; actually a direct connection is established.

> Windows Registry & User Profiles

During time the **Registry** of a Windows Operating system is filled with invalid or unnecessary entries. This could cause, that files are loaded into RAM or processes are started which are not required, thus just wasting system resources.
If you experience a difference in performance, depending on which user is logged on the same PC, this could indicate a Registry/**User Profile** problem.
To fix this, the User Profile should be deleted and re-created.

> 32bit Applications on 64bit Servers

It is not recommended to run 32bit applications on 64bit servers – at least not permanently or frequently – as this could cause degraded performance, or at worst an "application hang" or "server freeze".

The NAV/SQL Performance Field Guide
Version 2009

> Dynamic Management Views

With SQL Server 2005 the *"Dynamic Management Views"* (DMV) were introduced. These views are replacing the former "system tables"; providing lots of useful information.

The following excerpt lists several important DMV:

DMV	Description
sys.dm_db_index_physical_stats	Statistics about index size and fragmentation
sys.dm_db_index_usage_stats	Statistics about index usage
sys.dm_db_missing_index_details	Potentially missing indexes
sys.dm_exec_query_stats	Statistics about cached execution plans
sys.dm_exec_requests	Process/Session Information
sys.dm_exec_sessions	Process/Session Information
sys.dm_os_performance_counters	Performance counters and values
sys.dm_os_wait_stats	Statistics about system waits
sys.dm_os_sys_info	Various System/OS Information

Refer to the *Books Online* for further details.

STRYK System Improvement
Performance Optimization & Troubleshooting

Additional Resources

Here one could find further and detailed information:

Microsoft

http://www.microsoft.com/	Microsoft Homepage
http://www.microsoft.com/dynamics/nav/	Microsoft Dynamics NAV Homepage
http://www.microsoft.com/sql/	Microsoft SQL Server Homepage
http://support.microsoft.com/search/	Microsoft Support Knowledgebase
http://msdn2.microsoft.com/en-us/sql/	MS SQL Server at MSDN

Communities

http://dynamicsusers.net/	Dynamics User Group
http://www.mibuso.com/	Microsoft Business Solutions Online Community
http://www.msdynamics.de/	German MS Dynamics Community
http://www.sqlpass.org/	Professional Association for SQL Server
http://www.sql-server-performance.com/	MS SQL Server Community
http://www.sqlservercentral.com/	MS SQL Server Community

BLOGs

http://dynamicsusers.net/blog/stryk/	Jörg Stryk's BLOG at Dynamics User Group
http://blogs.msdn.com/microsoft_dynamics_nav_team_blog/	MS Dynamics NAV Team BLOG
http://blogs.msdn.com/microsoft_dynamics_nav_sustained_engineering/	MS Dynamics NAV Sustained Engineering Team BLOG
http://blogs.msdn.com/german_nav_developer/	German MS Dynamics NAV Development Team
http://blogs.msdn.com/mssqlisv/	Microsoft SQLCAT ISV Program Management Team

The NAV/SQL Performance Field Guide
Version 2009

Index

ndoar$ 60
ndodbconfig 116, 126, 144
ndodbproperty 62, 72, 178
ndoshadow 60

% Privileged Time 18
% Processor Time 18, 172

/PAE 43

1204 51, 132
1205 51, 132
1222 51, 132

3605 51, 132
3GB 43

4119 51, 117, 181
4616 51

845 46

Active/Active Cluster 146
Active/Passive Cluster 146
Activity Monitor 127
ADD CONSTRAINT 74, 93, 122
ADD FILE 85, 92
ADD FILEGROUP 85, 92
Address Windowing
  Extensions 44
ad-hoc 143
ALTER TABLE 74, 122
ALTER COLUMN 83
alter database 38, 39, 62
ALTER DATABASE 85, 92
ALTER INDEX 82
ALTER INDEX REBUILD 82
ALTER INDEX
  REORGANIZE 82
ALTER TABLE 83, 93
Always Rowlock 59, 118, 125, 126
application hang 129
Application Role 60
AS SNAPSHOT OF 149
Asynchronous Mirroring 50, 147, 148
Auto Growth 63

Auto. Close 55, 63
Auto. Create Statistics 55, 83
Auto. Growth 54, 63
Auto. Shrink 55
Auto. Statistics 63
Auto. Update Statistics 55
Auto. Update Statistics
  Asynchronously 55
AUTO_CLOSE 62
AUTO_CREATE_STATISTI
  CS 62
AUTO_SHRINK OFF 62
AUTO_UPDATE_STATISTI
  CS 62
AUTO_UPDATE_STATISTI
  CS_ASYNC 62
AutoIncrement 124
Automation Server 106
Available MBytes 17, 172
Avg. Disk Read Queue
  Length 17
Avg. Disk Sec/Transfer 17
Avg. Disk Write Queue
  Length 17
Avg. Read Queue Length 172
Avg. Write Queue Length 172
AWE 44
awe enabled 44, 61
AWE Enabled 52

Backup Compression 50
backup strategy 140
BCD 43
bcdedit.exe 43
BEGIN TRANSACTION 125
bigint 109
BigInteger 109
Binary 109
BLOB 109
Blocks 118
Bookmark Lookup 26
Boolean 109
Boot Configuration Data 43
Bottlenecks 15
B-Tree 65
buckets 89
Buffer Cache Hit Ratio 19, 173

C/AL 114
C/SIDE 114
C/SIDE Version 114
Checksum 55
Client Hardware
  Environment 40, 41, 42
Client Monitor 23
Clustered 71, 74
Clustered Index 65, 66
Clustered Index Scan 26
Code 109
Code Coverage 23
COMMIT TRANSACTION 125
Configuration 51
Context Switches/sec 18, 172
Costs per Record 69
Costs Per Record 89
COUNT 103
COUNT_BIG 95
COUNTAPPROX 103
Counter Log 22
covering index 90
Covering Index 98
CPU 29, 30, 40, 41, 42
CREATE DATABASE 149
CREATE INDEX 90, 93
CREATE NONCLUSTERED
  INDEX 74, 76, 80, 86, 99
CREATE UNIQUE
  CLUSTERED INDEX 80, 86, 122
CREATE UNIQUE
  NONCLUSTERED INDEX 76
CREATE VIEW 94
Cursor 104
Cursor Handling 72
Cursor Preparation 72

Data Compression 50
Data File(s) Size (KB) 20, 173
data integrity 148
Database 51
Database files 32
Database Mirroring 147, 149
Database Settings 54
Database Snapshots 149

Page 155

STRYK System Improvement
Performance Optimization & Troubleshooting

databaseversionno	178	Edition 43, 49	Hot Standby 146
database-versions	91	Empty SIFT Record 91	Hyperthreading 47
Data-Types	109	Enable for MS Dynamics	
Date	109	NAV Server 57	I/O Affinity 52
DateFormula	109	Error Log 132, 133, 137	IDENTITY INSERT 124
datetime	109	Exclusive Lock 119	IGNORE_DUP_KEY 87
DateTime	109	Exclusive Locks 119	image 109
DBCC CHECKDB	137	Execution Plans 25	implicit locking 131
DBCC DBREINDEX	82	explicit locking 131	INCLUDE 99
DBCC FREEPROCCACHE 145		Explicit Locking 119	Included Columns 67, 98, 99
DBCC INDEXDEFRAG	82		IncreaseUserVA 43
dbcc loginfo	20, 54	Failover Cluster 146	Index 65
DBCC MEMORYSTATUS 115		failover solutions 140	Index Hinting 117
DBCC PINTABLE	115	Fast Forward Cursor 142	Index Nodes 82
DBCC SHOW_STATISTICS 70		Fast Forward Cursors 71	Index Scan 26
DBCC SHOWCONTIG 79, 81		File-Group 85, 87, 92	Index Seek 26
DBCC TRACEOFF	132	Fill-Factor 78	Index Statistics 83
DBCC TRACEON	117, 132	FILLFACTOR 87	Indexed View 94
DBCC TRACESTATUS	132	Fill-Factors 77	Indexed Views 50
DBCC UNPINTABLE	115	Filter 26	Indexes 65
Deadlock	53	FILTERGROUPS 101	IndexHint 116
DEADLOCK_PRIORITY	129	FIND 101	indexproperty 56, 84
Deadlocks	118, 129	Find As You Type 57	Initial Size 54
decimal	109	FIND('-') 60, 102	INRAM 115
Decimal	109	FIND('-') 104	INSERT 131
DefaultLockGranularity	126	FIND('+') 102	int 109
DELETE	131	FINDFIRST 102, 104	Integer 109
DELETE FROM	91	FINDLAST 104	intended exclusive locks 125
diagnostics	72	FINDSET 60, 102, 104	Intended Exclusive Locks 119
Differential Backup	140	FINDSET(TRUE) 102, 104, 119	IsAutoStatistics 56, 84
Dirty Reads	119	FINDSET(TRUE, TRUE) 104	ISEMPTY 103, 104
Disk Subsystem	29, 30, 40, 41, 42	Flowfield 88	ISQLW 105
DisplaySingleLogSampleValue	22	fn_trace_getinfo 134	IsStatistics 56, 84
DMV	153	fn_trace_gettable 25	Key 65, 75
DROP CONSTRAINT 74, 93, 122		Foreign Key 65	
DROP INDEX	74	fragmentation 81	Leaf Nodes 82
DROP STATISTICS	56, 84	Free Pages 19	Light 28
DROP_EXISTING	87	Full Backup 140	Lightweightpooling 53
Dual CPU	47	Full Scans/sec 18, 173	LIKE 117
DualCore	47		Linked Object 108
Duration	25, 109	GHOST CLEANUP 152	Linked Server 108
Dynamic Cursor	142	gpedit.msc 46	load balancing 146
Dynamic Cursors	71	Group Policy 46	Lock
Dynamic Management View	22	GUID 109, 122	Deadlock 133
Dynamic Management Views	153	hardrowlock 62	Deadlock Chain 133
		Heavy 28	Deadlock Graph 133
		High Performance Mirroring 147	Lock Escalation 118
		High Security Mirroring 147	Lock Pages in Memory 44, 46
		HOLDLOCK 125	Lock Request/sec 19, 173
			Lock Timeout 59

The NAV/SQL Performance Field Guide
Version 2009

Lock Waits/sec	19, 173	
lock-escalation		125
Locking Mechanisms		59
Locking Order		131
Locking Order Rules		131
Locks		118
LOCKTABLE	103, 118, 131	
locktimeout		62
LOCKTIMEOUT		59
lodctr.exe		22
Log file(s) Size (KB)	20, 173	
Log Growths	20, 173	
Log Shipping	148, 150	
logical processors		47
LUN		35
Maintain		90
Maintain Defaults		58
Maintain Relationship		58
Maintain Views		58
MaintainSIFTIndex		89
MaintainSQLIndex		69
Maintenance		135
Maintenance Plan	80, 135	
Maintenance Plans		55
master	32, 37	
max server memory (MB)	61	
Max. Degree of Parallelism		53
Max. Server Memory		52
Max. Worker Threads		52
MAXDOP		87
Medium		28
Memory	17, 172	
Memory Grants Pending	19	
Merge Replication		150
Microsoft SQL Server 2000		49
Microsoft SQL Server 2005		49
Min. Server Memory		52
mirror		147
Mirroring		150
model	32, 38	
MODIFY		131
modify file	38, 39	
MOVE TO		93
MS ADO	106, 125	
msdb	32, 38	
mssqlsystemresource		32
MsSqlSystemResource		37
MTBF		33
Named Pipes		152
NAS		33
NAV Client Settings		57
NAV Database Sizing Tool		79
Network	29, 30, 40, 41, 42	
Network Interface	18, 173	
NOCHECK		93
Non-Clustered Index		66
Non-Clustered Indexes		65
NORECOVERY		147
Northwind		32
NT Fiber Mode		53
Number of Deadlocks/sec	19, 173	
objectproperty	56, 84, 115	
OLE-DB		106
ONLINE		87
Online Indexing		50
Operating System	29, 30, 40, 41, 42, 43	
Optimistic Concurrency	118, 120	
Option		109
OPTION(RECOMPILE)	117, 181	
ORDER BY	71, 101, 102	
ORDER BY DESC		102
order of fields		70
OSQL		105
Output Queue Length	18, 173	
Overflow Pages		54
PAD_INDEX		87
PAE		43
PAG IX		121
Page Life Expectancy		19
Page Split		77
Page Splits/sec	19, 173	
Page Verify		55
PAGE_VERIFY		62
Pages/sec	17, 172	
parameter sniffing		142
Percent Log Used	20, 173	
Performance		12
Performance Monitor		133
Performance Troubleshooting		13
Physical Address Extension		43
Physical Disk	17, 172	
physical processor		47
Plan Freezing		145
Plan Guides	117, 145	
Potential Deadlocks		131
Primary Key		65
PRIMARY KEY CLUSTERED		93
principal		147
Priority Boost		53
Processor	18, 172	
Processor Affinity		52
Processor Queue		47
Processor Queue Length	18, 172	
Pubs		32
Query Execution Plan		71
Querying SQL Server		101
quickfind		62
RAID		31
RAID 0		31
RAID 1		31
RAID 10		31
RAID 5		31
RAM	29, 30, 40, 41, 42	
Read Ahead pages/sec	173	
Reads	25, 99	
READUNCOMMITTED	102, 119	
RecompileMode		144
reconfigure	44, 61	
Record Set	40, 41, 60	
RecordID		109
RECORDLEVELLOCKING		118
RECOVERY		62
Recovery Model		63
Registry		152
REPAIR_FAST		137
REPAIR_REBUILD		137
REPEAT … UNTIL		104
Replication		150
Resource Governor		50
Response-Time		12
ROWLOCK		59
Row-Locking		125
SAN	33, 141	
SAS		33
SATA		33
SCHEMABINDING		95

STRYK System Improvement
Performance Optimization & Troubleshooting

SCSI		33	sp_trace_create		134	System	18, 172
Security Model		60	sp_trace_setevent		134		
SELECT		102	sp_trace_setstatus		134	–T4119	117
SELECT COUNT		103	sp_updatestats		83, 137	-T845	46
SELECT TOP 1		102	sp_who2		127	Table Optimizer	116
SELECT TOP 1 NULL		103	SQL Cursor		101	Table Scan	26
SELECT TOP 500		102	SQL Index	71, 74, 75, 142		TableFilter	109
selectivity		75	SQL Profiler		24, 133	TableIsPinned	115
Selectivity		70	SQL Server		49	TableView	101
semaphore		120	SQL Server Access Methods			Target Server Memory (KB)	
Semaphore		130			18, 173		19, 173
SERIALIZABLE		119	SQL Server Buffer Manager			TCP/IP	152
Serialization		130			19, 173	Technical Presales Advisory	
Server Hardware			SQL Server Databases		20,	Group	30
Environment		28	173			tempdb	32, 39, 63
server-site processing		105	SQL Server General			TempFilePath	40, 41
SET SHOWPLAN_ALL ON			Statistics		20, 173	temporary tables	112
		103	SQL Server Instance			Text	109
SET TRANSACTION			Settings		51	thread pooling	52
ISOLATION LEVEL			SQL Server Locks		19, 173	Throughput	12
SERIALIZABLE		102	SQL Server Memory			Time	109
SETCURRENTKEY		101	Manager		19, 173	Time %	17
SETFILTER		101	SQLCMD		105	tinyint	109
SETRANGE		101	sqlctr.ini		22	TO FILEGROUP	85, 92
Shared Locks		119	sqlctr80.dll		22	Torn Page Detection	55
show advanced options		44	sqlmaint.exe		135	TORN_PAGE_DETECTION	
SIFT		99	SQLOLEDB		106		62
SIFT Index		88	SSD		34	Total Server Memory (KB)	
SIFT Indexes	64, 88, 94		Standard Fill-Factor		53		19, 173
SIFT Table		88	STANDBY		148	TPAG	30
SIFTLevelsToMaintain		89	Startup parameter		51	Trace Flag	46
Single User		60, 137	Startup Parameter		132	Trace Flags	132
Snapshot Replication		150	Startup-Parameter		117	trace-flag	117
Snapshots		149	Streamlining		75	TRANSACTION ISOLATION	
Solid State Disks		34	Sum Index Flowfield			LEVEL	119
Sort		26	Technology		88	Transaction Log	148
SORT_IN_TEMPDB		87	SumIndexField		88	Transaction Log Backup	140
SourceTableTemporary		113	Synchronous Mirroring		147	Transaction Log files	32
sp_attach_db		39	sys.dm_db_index_physical_			Transaction Log Shipping	
sp_configure		44, 61	stats		79, 81, 153		148, 149
sp_control_plan_guide		145	sys.dm_db_index_usage_st			Transaction Speed	131
sp_create_plan_guide		145	ats		69, 153	Transactional Replication	150
sp_createstats		83, 137	sys.dm_db_missing_index_d			Transactions/sec	20
sp_cursoropen		72	etails		153	Transfer/Sec	17
sp_cursorprepare		72	sys.dm_exec_query_stats			Trigger	89
sp_cursorunprepare		72			153	TRN	148
sp_cycle_errorlog		137	sys.dm_exec_requests		153	Tuning Advisor	25
sp_delete_backuphistory		139	sys.dm_exec_sessions		153		
sp_detach_db		39	sys.dm_os_performance_co			U320	33
sp_helpfile		37	unters		22, 153	UNIQUE	68, 75
sp_helpindex		69	sys.dm_os_sys_info		153	uniqueidentifier	109
sp_helpstats		83	sys.dm_os_wait_stats		153	UPDLOCK	102
sp_spaceused		79	sys.plan_guides		145	User Connections	20, 173

User Profile	152	Virtual Logs	20, 54	Windows Performance	
UseRecompileForTable	144	Virtual Transaction Log	54	Monitor	16
User-Synchronization	60	VLog	20, 54	Windows Server 2008	43
		VSIFT	91, 94, 99	Winner	129
varbinary	109			WITH SCHEMABINDING	94
varchar	109	Warm Standby	147, 148, 150	Writes	25
Version	43, 49	WHERE	71, 101		
Version Principle	118	WHILE … DO	104	XDL	133
Victim	129	wildcards	101	XML	133
View	108	Windows Authentication	53	xp_cmdshell	61
Violation	131			xp_readerrorlog	132

STRYK System Improvement
Performance Optimization & Troubleshooting

Appendix A – System Checklists

Part A - SQL Server Configuration

Company Name: _____
Name: _____
Date: _____
Place: _____

Database Server General	
Description	
UNC Name	
IP Address	
Place	
Installation Date	
Administrator (E-Mail)	
Operating System	
Architecture	O x86 O x64 O i64
Comment	

The NAV/SQL Performance Field Guide
Version 2009

Hardware Specifications	
Number of CPU	
CPU Specification	
Memory Type	
RAM	
Network	
Hyper-Threading	O Yes O No
PAE	O Yes O No (with 32bit only)
3GB	O Yes O No (with 32bit only)
Comment	

Remarks:

_____

_____

_____

_____

_____

_____

_____

_____

_____

_____

_____

_____

STRYK System Improvement
Performance Optimization & Troubleshooting

Disk Subsystem	
Description	
Disk 1	
Drive Letter	
Controller Type	
RAID Configuration	
Capacity	
RPM	
Comment/Usage	
Disk 2	
Drive Letter	
Controller Type	
RAID Configuration	
Capacity	
RPM	
Comment/Usage	
Disk 3	
Drive Letter	
Controller Type	
RAID Configuration	
Capacity	
RPM	
Comment/Usage	

The NAV/SQL Performance Field Guide
Version 2009

Disk 4	
Drive Letter	
Controller Type	
RAID Configuration	
Capacity	
RPM	
Comment/Usage	

Disk 5	
Drive Letter	
Controller Type	
RAID Configuration	
Capacity	
RPM	
Comment/Usage	

Remarks:

_____
_____
_____
_____
_____
_____
_____
_____
_____

STRYK System Improvement
Performance Optimization & Troubleshooting

SQL Server Settings	
SQL Server Edition	
SQL Server Build No.	
Collation	
Affinity Mask	
CPU used	
Degree of Parallelism	
Threshold Parallelism	
AWE Enabled	O Yes O No (with 32bit only)
Max. Worker-Threads	
Memory Dynamic	O Yes O No
Memory Min.	
Memory Max.	
Authentication	O Mixed O Windows
Trace Flags	

Part B – Database Configuration

NAV Database Settings	
Current Database Size	GB:
Concurrent User	
NAV Build No.	
NAV DB Version No.	
Recovery Model	O **Full** O Simple O Bulk
Collation	
Auto. Close	O Yes O **No**
Auto. Shrink	O Yes O **No**
Auto. Create Stats	O Yes O **No**
Auto. Update Stats	O Yes O **No** O Asynchronously
Torn Page Detection	O Yes O **No**
Page Verify	O Torn Page Detection O **Checksum**
Maintain Defaults	O Yes O No
Maintain Views	O Yes O **No**
Maintain Relations	O Yes O **No**
Always Rowlock	O Yes O **No** ("Yes" with 64bit only)
Lock Timeout	O Yes O No Milliseconds:
Find As You Type	O Yes O No
Caching Record Set	
Security Model	O **Standard** O Enhanced

Remarks:

_____

_____

_____

STRYK System Improvement
Performance Optimization & Troubleshooting

NAV Database Settings	
File-Group(s) System	PRIMARY
File-Group(s) Data	Data Filegroup 1
Other File-Group(s)	
Auto. Growth Data	O MB O % Amount:
Auto. Growth Log	O MB O % Amount:

Remarks:

___
___
___
___
___
___
___
___
___
___
___

```sql
-- EXECUTE WITHIN CONTEXT OF NAV-DATABASE --

select @@servername as "ServerName", @@servicename as "ServiceName",
@@language as "Language", @@version as "ServerVersion"
go
select * from sys.dm_os_sys_info
go
sp_configure 'show advanced options', '1'
reconfigure
go
sp_configure
go
select db_id() as "DatabaseID", db_name() as "DatabaseName"
go
select * from sysfiles
go
-- see "Books Online": DATABASEPROPERTYEX
select DATABASEPROPERTYEX(db_name(), 'Collation') as 'Collation'
select DATABASEPROPERTYEX(db_name(), 'IsAutoClose') as 'IsAutoClose'
select DATABASEPROPERTYEX(db_name(), 'IsAutoCreateStatistics') as
'IsAutoCreateStatistics'
select DATABASEPROPERTYEX(db_name(), 'IsAutoUpdateStatistics') as
'IsAutoUpdateStatistics'
select DATABASEPROPERTYEX(db_name(), 'IsAutoShrink') as 'IsAutoShrink'
select DATABASEPROPERTYEX(db_name(), 'IsTornPageDetectionEnabled' ) as
'IsTornPageDetectionEnabled'
select DATABASEPROPERTYEX(db_name(), 'Recovery') as 'Recovery'
select DATABASEPROPERTYEX(db_name(), 'Status') as 'Status'
select DATABASEPROPERTYEX(db_name(), 'UserAccess') as 'UserAccess'
select DATABASEPROPERTYEX(db_name(), 'Version') as 'DBVersion'
go
select "databaseversionno", "maintainviews", "maintainrelationships",
"checkcodepage", "quickfind", "maintaindefaults", "locktimeout",
"locktimeoutperiod", "hardrowlock", "bufferedrows", "securityoption"
from "$ndo$dbproperty"
go
select * from "$ndo$dbconfig"
go
select object_name([id]) as [Heap Tables] from sysindexes
where indid = 0
go
```

STRYK System Improvement
Performance Optimization & Troubleshooting

Part C – Database Maintenance

Maintenance Jobs	
Cycle Error Logs	O Yes O No O Name:
Keep Logs	
Maintain Stats	**O Yes** O No O Name:
Maintain Indexes	**O Yes** O No O Name:
SIFT Maintenance	O Yes O No O Name:
Clean Up History	O Yes O No O Name:
Deadlock Detection	**O Yes** O No O Name:
Block Detection	**O Yes** O No O Name:
Integrity Check	**O Yes** O No O Name:

Remarks:

_____
_____
_____
_____
_____
_____
_____
_____

The NAV/SQL Performance Field Guide
Version 2009

Part D – Performance Monitor

Measurement Start Date/Time: _____

Measurement End Date/Time: _____

Description: _____

Performance Monitor

Object	Counter	Instance	Best	Avg.
Memory	Available MBytes	n/a	> 5	
Memory	Pages/sec	n/a	< 25	
Physical Disk	Avg. Read Queue Length	NAV Database file	< 2	
Physical Disk	Avg. Read Queue Length	NAV Transaction Log file	< 2	
Physical Disk	Avg. Write Queue Length	NAV Database file	< 2	
Physical Disk	Avg. Write Queue Length	NAV Transaction Log file	< 2	
Physical Disk	Time %	NAV Database file	< 50	
Physical Disk	Time %	NAV Transaction Log file	< 50	
Physical Disk	Avg. sec/Transfer	NAV Database file	< 0,015	
Physical Disk	Avg. sec/Transfer	NAV Transaction Log file	< 0,015	
Physical Disk	Transfers/sec	NAV Database file	< 120	
Physical Disk	Transfers/sec	NAV Transaction Log file	< 120	
Processor	% Processor Time	Total	< 80	
Processor	% Privileged Time	Total	< 10	
System	Processor Queue Length		< 2	
System	Context Switches/sec		< 8000 per CPU	

STRYK System Improvement
Performance Optimization & Troubleshooting

Object	Counter	Instance	Best	Avg.
Network Interface	Current Bandwidth		> = 1GB	
Network Interface	Output Queue Length		< 2	
SQL Server Access Methods	Full Scans/sec	NAV database	n/a	
SQL Server Access Methods	Page Splits/sec	NAV database	n/a	
SQL Server Buffer Manager	Buffer Cache Hit Ratio		> 90	
SQL Server Buffer Manager	Free Pages		> 640	
SQL Server Buffer Manager	Page Life Expectancy		> 300	
SQL Server Locks	Lock Request/sec	Total	n/a	
SQL Server Locks	Lock Waits/sec	Total	0	
SQL Server Locks	Number of Deadlocks/sec	Total	0	
SQL Server Databases	Data File(s) Size (KB)	NAV database	n/a	
SQL Server Databases	Log file(s) Size (KB)	NAV database	n/a	
SQL Server Databases	Log Growths	NAV database	0	
SQL Server Databases	Percent Log Used	NAV database	n/a	
SQL Server General Statistics	User Connections		n/a	
SQL Server General Statistics	Processes blocked		0	
SQL Server Memory Manager	Total Server Memory (KB)		n/a	
SQL Server Memory Manager	Target Server Memory (KB)		n/a	
SQL Server Memory Manager	Memory Grants Pending		0	

(use "*Report*" mode [display *average* figures])

The NAV/SQL Performance Field Guide
Version 2009

Part E – SQL Profiler

Measurement Start Date/Time: _____

Measurement End Date/Time: _____

Description: _____

SQL Profiler		
Events	**Columns**	**Filters**

Stored Procedures
  RPC: Completed
  SP: Completed
  SP: StmtCompleted
 	
TSQL
  SQL: BatchCompleted
  SQL: StmtCompleted
	Groups
	  <none>
	Columns
	  SPID
	  TextData
	  Reads
	  Writes
	  CPU
	  Duration
	  Start Time
	  End Time
	  LoginName
	  HostName
	  Application Name
	  EventClass
		Application Name
		  Not Like %SQL%
		
		Reads
		  Greater/Equal than 1000
		
		Duration
		  Greater/Equal than 30

Remarks:

_____

_____

_____

_____

_____

Appendix B – Version Lists

MS Dynamics NAV (Navision) C/SIDE Versions

Version	Build	KB Article
3.70.B	19516	
	19868	
4.00	19365	
	20942	
	21871	
4.00 SP1	21666	
	21990	
	22373	
	22363	
4.00 SP2	22100	
	22611	
	22851	
	23099	
	23460	
4.00 SP3	23305	
	24080	931841
	24219	933727
	24449	936602
	24734	938138
	25143	940718 [1]
	25202	940718
	25206	942540
	25265	934409
	25307	940643
	25321	943210
	25382	944204
	25484	944773
	25598	945264
	25638	945349
	25709	946247
	25726	945992

25732	945339
	946204
25906	947574
25994	948580
26033	948302
	942405
26170	948824
26410	950920
26535	952201
26565	952355 [2]
26708	952873
26752	953313
26954	929047
27010	954342
	954722
27256	957275
27371	957219
27742	959165
27765	959822
27857	959656
28477	965225
28541	967469
28909	962005
29021	970383
29113	971172 [2]
29444	973458
29539	968649
29649	974890
29689	943496
29723	974795
	975345

[1]) Minimum Requirement; Supported with Windows Server 2008 and SQL Server 2008
[2]) Minimum Recommended

21st Sept. 2009. This list may not be complete. Not all updates are relevant for NAV with SQL Server.
See also http://blogs.msdn.com/german_nav_developer/

The NAV/SQL Performance Field Guide
Version 2009

Version	Build	KB Article
5.00	24199	
	24632	
	25344	
	25359	
	25560	
	25581	
	25653	
	25684	
	26026	
	26810	
5.00 SP1	26084	954191
	26654	953245
	26751	951631
		952373
		953545
	26810	954342
	26948	954672
	27002	
	27191	956161 [1]
	27199	957276
	27241	957267
	27253	954722
	27368	957219
		957471
	27398	957824
	27831	959911
		959912
		960141
	28007	960607
	28028	960661
	28342	962005
	28412	963700

	28512	967158
	28563	967318
	28715	968354
	28798	968465
	28874	969037
	28884	959158
		968679
		969452
	28913	969590
	28956	969881
	29040	969777
	29048	970545
	29048	970545
	29083	970913
	29118	971172 [2]
	29125	971170
	29178	970256
	29245	971797
	29299	972431
	29379	972793
		973133
	29390	973043
	29410	973045
	29469	973626
	29497	973900
	29512	973954
	29550	974152
	29608	970502
	29715	975273
	29729	974798
	29736	974802
	29763	974523

[1]) Minimum Requirement; Supported with Windows Server 2008 and SQL Server 2008
[2]) Minimum Recommended

21st Sept. 2009. This list may not be complete. Not all updates are relevant for NAV with SQL Server. See also http://blogs.msdn.com/german_nav_developer/

STRYK System Improvement
Performance Optimization & Troubleshooting

Version	Build	KB Article
2009	26370	CTP3
	27205	CTP4
	27808	RTM
	27980	960268
	28177	961772
	28463	962012
	28622	961430
	28713	968196
	28766	968650
	28772	968466
	28795	968189
		968649
	28815	968277
	28829	968411
	28889	969447
	28981	969448
		969649
	29070	970473
	29082	970762
	29096	970712
	29140	970545
		971124
		971173
	29202	970079
	29210	971174
		971738
	29267	972059
	29285	971989

			972372
			972465
			972660
			971784
		29372	972430
			972791
			972832
		29392	972970
			973041
			942540
		29415	971519
			972522
			973237
			973227
		29443	973416
			973491
		29500	973899
			973959
		29515	973997
		29518	974024
		29532	974153
		29548	974196
		29607	974526
		29683	974244
		29735	974718
		29762	975799
2009 SP1		29626	SP
		29741	974980
		29766	974798

21st Sept. 2009. This list may not be complete. Not all updates are relevant for NAV with SQL Server. See also http://blogs.msdn.com/german_nav_developer/

The NAV/SQL Performance Field Guide
Version 2009

MS Dynamics NAV (Navision) Database Versions

"**databaseversionno**" as stored in table "**ndodbproperty**".
View number using TSQL or Management Studio/Enterprise Manager:

```
USE [Navision]
GO
SELECT [databaseversionno] FROM [dbo].[$ndo$dbproperty]
```

NAV Version	DB Version	Description
2.50	1	Beta 1 data formats
	2	Beta 2 data formats (change in Code field value format from space-padded prefix to hex prefix)
	3	Beta 3 data formats (change in Code field value format from hex prefix to no prefix or special format)
	4	Release data formats (SIFT trigger identifiers changed and new indexes on SIFT tables)
2.60 (A-C)	5	Release data formats (change in Session view and SIFT triggers to use CONVERT rather than CAST for datetime fields)
2.60 (D-F)	6	Changes in SIFT triggers due to UPDATE... WHERE... performance problem
3.00	7	Beta 1 data formats (change in Session view for SQL Server 2000; change in chartable)
	8	Beta 1 data formats (change in chartable; changes in SIFT triggers due to UPDATE... WHERE... performance problem)
	9	Beta 1 data formats (added maintainviews dbproperty)
	10	Beta 2 data formats (mapped invalid characters to "_" for SQL object identifiers)
	11	Release data formats (added diagnostics dbproperty)
3.01 (A-B)	12	Release data formats (added identifiers dbproperty)
3.10 (A)	13	Release data formats (added maintain relationships dbproperty)
3.60	14	Release data formats (change in Session view for new extensions; identifier conversion dbproperties; per-database license dbproperty)

STRYK System Improvement
Performance Optimization & Troubleshooting

NAV Version	DB Version	Description
3.70	15	Release data formats (change in chartable; added check codepage, quick find, maintain defaults dbproperties; change in Session view for new extensions and column COLLATE; change in Database File view for column COLLATE)
	16	Hotfix 5 data formats (change in Session View for removing duplicate connections based on 'sysprocesses.ecid')
	17	Hotfix 12 data formats (change in Session View for removing duplicate connections based on 'sysprocesses.ecid'; the identifiers and invalid identifier-chars dbproperties are updated to OEM)
3.70 (B)	18	data formats (change in Session View, removing join to syslockinfo; also removed from session count query)
4.0	20	Pre-release data formats (change in Session view for new Idle Time column; change in permission table to include new object types)
	30	Security release data formats (change in Session view, removing join to syslockinfo; change in chartable; creation of security objects)
4.1	40	Release data formats (added lock timeout, lock timeout period, hard rowlock and buffered rows dbproperties)
4.2	50	Update 3 (no actual database conversion was made)
4.2 U4 4.3	60	Alterable security option; fix for SIFT data corruption around Closing Dates (added security option dbproperty)
	61	Rebuilding SIFT triggers in order to correct the Sum problem
	62	Rebuilding SIFT triggers in order to correct the delete statement
	63	Prevention of updating SIFT twice
5.0	80	Release data formats (added system tables for office integration and record links; change in table descriptions for clustered property)
	81	Rebuilding SIFT triggers in order to correct the Sum problem
	82	Rebuilding SIFT triggers in order to correct the delete statement
5.0 SP1	95	Removed SIFT tables and triggers, creating Indexed Views instead
6.0	120	Release data formats (incl. CTP1) (added several new system tables; extended record links system table)
	130	Pre-release data formats (incl. CTP2) (removed SIFT tables and triggers, creating Indexed Views instead)
	140	Pre-release data formats (removed Assembly and Relationship system tables; added enabled for service dbproperty)

[Taken from a reply on my BLOG]

The NAV/SQL Performance Field Guide
Version 2009

MS SQL Server 2000 Versions

Description	Build
RTM	8.00.194
SP1	8.00.384
SP2	8.00.534
SP3	8.00.760
SP4	8.00.2039
Cumulative Update	8.00.2040
Cumulative Update	8.00.2145
Cumulative Update	8.00.2146
Cumulative Update	8.00.2148
Cumulative Update	8.00.2151
Cumulative Update	8.00.2156
Cumulative Update	8.00.2159
Cumulative Update	8.00.2162
Cumulative Update	8.00.2164
Cumulative Update	8.00.2166
Cumulative Update	8.00.2171
Cumulative Update	8.00.2172
Cumulative Update	8.00.2175
Cumulative Update	8.00.2177
Cumulative Update	8.00.2180
Cumulative Update	8.00.2181
Cumulative Update	8.00.2184
Cumulative Update	8.00.2185
Cumulative Update	8.00.2189
Cumulative Update	8.00.2190
Cumulative Update	8.00.2191
Cumulative Update	8.00.2192
Cumulative Update	8.00.2194

Description	Build
Cumulative Update	8.00.2196
Cumulative Update	8.00.2197
Cumulative Update	8.00.2199
Cumulative Update	8.00.2200
Cumulative Update	8.00.2201
Cumulative Update	8.00.2203
Cumulative Update	8.00.2204
Cumulative Update	8.00.2207
Cumulative Update	8.00.2209
Cumulative Update	8.00.2215
Cumulative Update	8.00.2216
Cumulative Update	8.00.2218
Cumulative Update	8.00.2220
Cumulative Update	8.00.2223
Cumulative Update	8.00.2226
Cumulative Update	8.00.2231
Cumulative Update	8.00.2232
Cumulative Update	8.00.2234
Cumulative Update	8.00.2238
Cumulative Update	8.00.2239
Cumulative Update	8.00.2241
Cumulative Update	8.00.2244
Cumulative Update	8.00.2245
Cumulative Update	8.00.2246
Cumulative Update	8.00.2248
Cumulative Update	8.00.2249

24th Sept. 08. This list may not be complete.

See http://support.microsoft.com/kb/894905 for details.

STRYK System Improvement
Performance Optimization & Troubleshooting

MS SQL Server 2005 Versions

Description	Build
RTM	9.0.1399
SP1	9.0.2047
SP2	9.0.3042
SP2a	9.0.3042.01
Cumulative Update	9.0.3152 [1]
Cumulative Update	9.0.3161
Cumulative Update	9.0.3175
Cumulative Update	9.0.3186
Cumulative Update	9.0.3200 [2]
Cumulative Update	9.0.3215
Cumulative Update	9.0.3228
Cumulative Update	9.0.3239
Cumulative Update	9.0.3257
Cumulative Update	9.0.3282
Cumulative Update	9.0.3294
Cumulative Update	9.0.3301
Cumulative Update	9.0.3315
Cumulative Update	9.0.3325
Cumulative Update	9.0.3328
Cumulative Update	9.0.3330
SP3	9.0.4035
Cumulative Update	9.0.4207
Cumulative Update	9.0.4211
Cumulative Update	9.0.4220
Cumulative Update	9.0.4226
Cumulative Update	9.0.4230

21st Sept 09. This list may not be complete.

See http://support.microsoft.com/kb/937137/ and
http://support.microsoft.com/kb/960598/ for details.

[1]) Minimum requirement to use the **OPTION (RECOMPILE)** features
[2]) Required to use new trace-flag **4119**

MS SQL Server 2008 Versions

Description	Build
RTM	10.0.1600
Cumulative Update	10.0.1763
Cumulative Update	10.0.1779
Cumulative Update	10.0.1787
Cumulative Update	10.0.1798
Cumulative Update	10.0.1806
Cumulative Update	10.0.1812
Cumulative Update	10.0.1818
SP1	10.0.2531
Cumulative Update	10.0.2710
Cumulative Update	10.0.2714
Cumulative Update	10.0.2723

21st Sept 09. This list may not be complete.

See http://support.microsoft.com/kb/956909/ for details.

STRYK System Improvement
Performance Optimization & Troubleshooting

"*The NAV/SQL Performance Field Guide*" is a booklet for the experienced NAV/SQL administrator which explains in a simple and easily understandable way the known (and unknown?) performance problems encountering with *Microsoft® Dynamics™ NAV* (*Navision®*) and *Microsoft® SQL Server™* and gives practical advices, solutions and recommendations to fix and avoid problems; covering all major areas of performance optimization and troubleshooting:

- ✓ Performance Measurement & Analysis
- ✓ Platform Setup & Configuration
- ✓ SQL Server Configuration
- ✓ Database Setup
- ✓ Index- and SIFT/VSIFT Tuning
- ✓ Block- and Deadlock Detection
- ✓ Query Analysis
- ✓ C/AL™ Code Optimization
- ✓ and much more!

Applies to

Navision® Financials™ 2.60/2.65
Navision® Attain™ 3.01/3.60/3.70
Microsoft® Business Solutions Navision™ 4.00
Microsoft® Dynamics™ NAV 5.00
Microsoft® Dynamics™ NAV 5.00 SP1
Microsoft® Dynamics™ NAV 2009
Microsoft® Dynamics™ NAV 2009 SP1

with

Microsoft® SQL Server™ 2000
Microsoft® SQL Server™ 2005
Microsoft® SQL Server™ 2008

ISBN: 978-3-8370-1442-6